Entrepreneurship and Innovation During Austerity

Also by Ian Chaston (*Palgrave Macmillan)

VALUES DRIVEN PUBLIC SECTOR REFORMATION*

STRATEGIES FOR SUSTAINING COMPETITIVE ADVANTAGE: Surviving Declining Market Demand and China's Globalisation

MARKETING DICTIONARY, i-App Delivery

PUBLIC SECTOR MANAGEMENT*

ENTREPRENEURSHIP AND SMALL FIRMS

BOOMER MARKETING: Exploiting A Recession Resistant Consumer Group

KNOWLEDGE-BASED MARKETING

SMALL BUSINESS E-COMMERCE MANAGEMENT*

SMALL BUSINESS MARKETING*

E-COMMERCE MARKETING MANAGEMENT

ENTREPRENEURIAL MARKETING

Entrepreneurship and Innovation During Austerity

Surviving Beyond the Great Recession

Ian Chaston
Centrum Catolica, Peru and Moustraining Ltd, UK

© Ian Chaston 2013

All rights reserved. No reproduction, copy or transmission of this publication may be made without written permission.

No portion of this publication may be reproduced, copied or transmitted save with written permission or in accordance with the provisions of the Copyright, Designs and Patents Act 1988, or under the terms of any licence permitting limited copying issued by the Copyright Licensing Agency, Saffron House, 6–10 Kirby Street, London EC1N 8TS.

Any person who does any unauthorized act in relation to this publication may be liable to criminal prosecution and civil claims for damages.

The author has asserted his right to be identified as the author of this work in accordance with the Copyright, Designs and Patents Act 1988.

First published 2013 by
PALGRAVE MACMILLAN

Palgrave Macmillan in the UK is an imprint of Macmillan Publishers Limited, registered in England, company number 785998, of Houndmills, Basingstoke, Hampshire RG21 6XS.

Palgrave Macmillan in the US is a division of St Martin's Press LLC, 175 Fifth Avenue, New York, NY 10010.

Palgrave Macmillan is the global academic imprint of the above companies and has companies and representatives throughout the world.

Palgrave® and Macmillan® are registered trademarks in the United States, the United Kingdom, Europe and other countries.

ISBN 978–1–137–32442–9

This book is printed on paper suitable for recycling and made from fully managed and sustained forest sources. Logging, pulping and manufacturing processes are expected to conform to the environmental regulations of the country of origin.

A catalogue record for this book is available from the British Library.

A catalog record for this book is available from the Library of Congress.

Typeset by MPS Limited, Chennai, India.

Contents

List of Figures vi
Preface vii
1 The New Austerity 1
2 Customer Values 19
3 Revisiting Management Philosophies 37
4 Assessing Futures 55
5 Tomorrow's Markets 69
6 Tomorrow's Competences 87
7 Leadership, Vision and Strategy 105
8 Innovation Strategies 125
9 Technology Strategies 143
10 Strategy Implementation 161
11 The Service Sector 177
12 The B2B Sector 191
13 Small Firms 211
14 The Public Sector 225
15 Leaner Futures 241
Index 259

List of Figures

3.1	Marketing outcomes	38
3.2	Possible components in an entrepreneurial approach to planning for tomorrow's world	49
6.1	Entrepreneurship competence options	97
7.1	Mission and factors of influence	113
7.2	Customer product space map	114
7.3	Entrepreneurial strategy options	116
7.4	Strategic positioning options	118
8.1	Alternative innovation propositions	127
9.1	Factors of influence	144
9.2	Alternative market opportunities matrix	148
9.3	Product/market change matrix	150
9.4	The innovation S-curve	156
11.1	Diversified consumer market service gap model	182
15.1	Identifying smart strategy and structure	253

Preface

Chapter 1 examines the economic and political events which have led to the emergence of the new age of austerity. Effective management of customer buying behaviour requires a detailed understanding of prevailing personal and societal values. Chapter 2 examines the nature of customer values and how these have been influenced by the new austerity; thereby leading to a change in consumption patterns for certain products and services. Changing purchase behaviour raises the question of whether certain prevailing market management theories now remain applicable or require revision or replacement. Chapter 3 examines the relevance of certain existing management theories and reviews whether revisions may be required to enhance their relevance in the market conditions that now prevail in many Western economies.

Although the new austerity was triggered by the Great Recession and governments' need to reduce public sector spending, Western consumers' spending power will be further eroded by the existence of additional global meta-trends. Chapter 4 examines the nature of these trends in relation to factors such as rising energy costs, increasing food prices and population ageing. Consumers' response to reduced disposable incomes will be reflected in the emergence of new or revised consumption patterns. Chapter 5 examines the nature of these changing patterns and how the new austerity will alter consumer motivation and purchase behaviour.

As markets and customer behaviour change in the face of the new austerity, not all of the existing competences that have supported organisational performance in the past may be relevant in the future. Chapter 6 covers the issues of the competences required to exhibit a dynamic response to new external environments and the nature of the entrepreneurial and innovation competences that will permit organisations to develop more relevant products or revise their market positioning.

Chapter 7 examines the role of entrepreneurship as the basis of defining organisational objectives and determining the most effective strategy for achieving defined performance objectives in markets where customer behaviour has changed as a result of the new austerity. Converting entrepreneurial strategies into viable product or service propositions often requires radical innovation. Chapter 8 examines the processes associated with exploiting radical innovation in today's organisations. The comparative cost advantage enjoyed by nations such as China and India means firms located in these countries are more able to exploit lower prices as a strategy for gaining customer share in Western markets affected by the new austerity.

To ensure their survival, Western organisations will probably need to rely on the exploitation of advances in technology to counter this threat. Chapter 9 reviews how knowledge-based Western nations can utilise new technology as the basis for surviving through reliance upon entrepreneurship and innovation.

Strategies can only deliver enhanced performance through the successful implementation of proposed actions. Chapter 10 examines the processes associated with the effective implementation of strategies reliant upon innovation. The role and management of entrepreneurship is influenced by the market sectors and the nature of benefits offered to customers. Chapters 11, 12, 13 and 14, respectively, review the differing processes and policies that are required in service businesses, small firms, B2B markets and public sector organisations.

The advent of electronic convergence and increasingly powerful data management capabilities are providing opportunities based upon exploiting smart technology to develop new products and to enhance organisational productivity. Chapter 15 examines how smart technology offers Western organisations the prospect of renewed economic growth across a diversity of sectors, such as consumer services, supply chain operations and healthcare, by using smart technology to fulfil customer needs during the current period of austerity and beyond.

The primary aim is to provide an alternative text and further reading for undergraduate and postgraduate business and management programmes where students will require access to new ideas and thinking, following the severely reduced relevance of many existing management theories due to recent economic events and the subsequent onset of what is expected to be a prolonged period of austerity within many Western nations. A secondary audience is managers in those Western organisations who are struggling to sustain organisational performance during the current economic downturn and are seeking to identify new strategies to more effectively satisfy the needs of their customers whose buying behaviour has been altered by low priced imports from Asia, adverse economic events and government spending cuts.

1
The New Austerity

The golden age

Although the Industrial Revolution accelerated wealth generation in Western nations, only following the Great Depression and World War II did the quality of life for the majority of the people begin to show any real improvement (Marr, 2012). The first beneficiaries after World War II were citizens of the USA, which had become the world's wealthiest nation. As this country shifted from a war to a peacetime economy, jobs were plentiful and incomes began to rise. Europeans were less fortunate. Another 10 years would pass before their economies, assisted by America's Marshall Plan, began to recover from the devastation wrought by years of war. However even before their economies began to grow, most countries had introduced a more equitable welfare state offering free or subsidised healthcare and education. Furthermore the adoption of Keynesian economics, involving the use of public sector spending to promote economic growth, had a major impact on reducing unemployment.

During the 1950s and 1960s most Western economies enjoyed a golden age of high employment and rising incomes. This provided the basis for the creation of the world's first experience of mass consumption, in which the average citizen in Western nations enjoyed the benefits of being able to afford goods such as convenience foods, domestic electrical appliances and package holidays. Firms that benefited from the advent of mass consumption included Proctor & Gamble, Nestlé, Unilever, Coca Cola and General Foods.

The first cracks

One of the factors influencing per capita spending power in the 1950s and 1960s was the low price of oil, which provided the feed stocks for plastic goods, petrol for cars, low cost heating and highly affordable energy supplies to support the manufacture and transportation of goods. The low price of oil reflected the fact that the major oil companies had control over huge oil deposits in areas of the world such as the Middle East. Paying third world countries low royalties kept oil costs for Western democracies at only a few cents per barrel. Hence it can be argued that a key factor determining the scale of the post-war consumer boom was a long period of virtually zero energy costs.

Having failed to persuade Western oil companies to accept higher oil prices, Arab countries formed the Organization of the Petroleum Exporting Countries (OPEC) and, in the 1970s, initiated two global oil embargoes. This was the first indication that Western economies should expect an upward trend in energy prices. This trend has continued through to the present day and can be expected to have an ongoing adverse impact on economic growth in the future. By the 1970s Western nations were facing increasing levels of inflation. Although the activities of the OPEC nations provided an excellent scapegoat, the real major problem was that Western politicians had become totally enamoured by Keynesian economics. They had realised that increasing public sector expenditure would create more jobs, which greatly enhanced their prospects for being re-elected. Politicians seemed to ignore John Maynard Keynes' strong disapproval of long periods of deficit spending on social programmes as a means of stimulating consumption. In his view, deficits in social programme expenditure should only occur during an economic downturn and these deficits should be funded by the accumulation of public sector financial surpluses from taxes raised during periods of economic growth. Keynes was also opposed to policies aimed at varying incomes via tax policy in order to stimulate consumption. In his view, the outcome of such policies would be inflation, which in turn would eventually lead to even higher unemployment. By the 1980s, after several decades of deficit spending, this is exactly what occurred (Brown-Collier and Collier, 1995).

Adding to the problem of rising inflation, major Western firms needed to combat increasing wages in their domestic plants to compete against the new Tiger Nations such as Japan and Taiwan. Some firms relocated their manufacturing operations to countries in Asia with lower labour costs. In addition to exporting jobs, these Western firms also provided potential competitors in these overseas countries with access to their technological expertise. Initially firms in Asia merely acted as satellite suppliers of goods. However, as they gained understanding of Western manufacturing techniques and business strategies Asian companies began to supply own label goods to major retailers in Europe and the USA. From this bridgehead it was a relatively small step for firms such as Honda, Sony, Toyota and Toshiba to start marketing their own branded products in Western markets.

Rising unemployment and inflation caused the unions in some European countries to become extremely militant. This militancy was especially prevalent in the public sector, in part reflecting the fact that strikes by key public sector workers had the capability to achieve disruption on a national scale. A decline in the general public's support for the unions became apparent in a number of countries as the 1980s progressed. The UK's Conservative prime minister Margaret Thatcher was elected at the beginning of the 1980s on a platform of using privatisation to reduce the power of the unions across industries still in the public sector, such as coal, steel and utilities (Babcock, Engberg and Glazer, 1997).

By the mid-1980s, inflation and rising unemployment were problems confronting virtually every Western democracy. With any reduction in the size of the public sector deemed as unacceptable by politicians concerned about losing support among the electorate, governments under the auspices of New Public Management (NPM) began to draw upon managerial principles from the private sector in an attempt to make public sector organisations more effective and efficient (Kim and Hong, 2006). In the UK and the USA increasing concerns about the ongoing viability of funding the public sector saw Keynesian economic theory being challenged by neoclassical economists. These theorists expressed concern that the inevitable outcome of prolonged deficit spending would move beyond merely stimulating inflation to actually causing the collapse of economies.

One of the leading opponents of governments continuing to expand the size of public sector deficits through borrowing was the University of Chicago Professor Milton Friedman. Referred to as a 'monetarist', Friedman produced a number of academic papers and the hugely successful book entitled *Capitalism and Freedom* that sought to demonstrate the abuses which can be caused by the misapplication of Keynsian economic theories. His perspective on monetary theory was that, in order to defeat inflation, governments should use Central Banks to establish stable monetary policies. Concurrently the emphasis should be on creating an affordable welfare state by focusing on promoting the wealth-generation activities of capitalism, which could generate the taxes that provided a non-inflationary supply of public sector funds (Jordan et al, 1993).

During the period of popularity for monetarist economics, little mention was made of another, much earlier critic of Keynes, a member of the Austrian School of economics, Friedrich A. von Hayek (Bas, 2011). Keynes' argued that total aggregate demand determines employment and hence increased spending increases employment and an economic crisis is resolved. This concept has been used by governments for over 100 years to justify policies involving increased public sector spending. Hayek's criticism was that Keynes' model ignored the theory of capital, and that government spending to influence demand–employment relationships can only be effective over the short term. Hence the Keynesian philosophy of 'in the long run, we are all dead' was seen by Hayek as academic irresponsibility, because such thinking leads to policies which can be extremely harmful to a nation's economy over the longer term. Furthermore government spending often is targeted at the weakest industries. Hayek felt this created further distortion in the allocation of resources; thereby possibly triggering an economic boom, which inevitably will be followed by an economic bust.

A frequent reason given by other schools of economic thought for rejecting Hayek's view was that his thesis was not relevant to post-war events, because he was commenting on the applicability to a system of wholesale state planning and the eventual outcome for Keynesian policies being the emergence of an autocratic, dictatorship. In reality this perspective ignored the fact that Hayek was developing his ideas in the 1930s when some European countries were moving towards fascism as a response to the perceived growing threat of communism. By continuing to express such comments, Hayek's critics were clearly ignoring his subsequent up-dating of ideas and writings, which he produced for many years following the end of World War II.

By the 1990s politicians supported a monetary policy of Central Bankers keeping interest rates low to combat inflation. This policy also provided the benefit of reducing government borrowing costs. Although politicians were aware of the growing problems in servicing growing public sector debt, in Europe concerns about assisting the unemployed, the elderly and socially disadvantaged led to further increases in public sector spending. Total public sector spending in the developed nations increased from 12.6 percent of GDP in 1960 to 17.3 percent in 1995 (Tanzi and Schuknecht, 2000).

A recession in the early 1990s provided an early warning that a slowdown in economic growth across the Western democracies would reduce consumers' ever increasing ability to enjoy ongoing improvement in their standard of living. An actual decline in consumer wealth was averted by low interest rates in countries such as the UK and the USA, triggering an upward house price spiral. Owner-occupiers perceived the apparently never ending increase in the value of their homes as justification for more borrowing to fund their consumerist lifestyle. Concurrently those wanting to buy their first home were eager to accept loans from the banks willing to lend up to 120 percent of the value of their purchase. The banks' behaviour merely contributed to further rises in house prices and sustained home owners' use of this asset to borrow even more money (Whalen, 2008). By the end of the 1990s many consumers in Western democracies felt that their prospects for greater increases in their personal income, through further gains in the value of property, were extremely good. Concurrently increases in public sector expenditure led the unemployed and socially disadvantaged to also enjoy some ongoing improvement in their standard of living.

Elephants in the room

By the late 1980s politicians in those Western democracies where healthcare is provided either free or subsidised by the state were aware that advances in medical technology, when combined with longer life expectancies, meant this area of the public sector was becoming unaffordable. In those countries, such as the USA, where people or their employers pay for healthcare through private insurance but qualify for federally funded Medicare upon retirement, these rising costs were also recognised as being economically unsustainable. Although politicians introduced public sector healthcare reforms, very few of these actions had any real positive impact on service provision efficiency. Hence the total cost of government spending on healthcare continued to rise (Chaston, 2011).

Politicians ignored the need for spending reforms as response to the problems associated with population ageing (Johnson, 2004). Population ageing is the term used to describe the effect of rising average age and older people becoming the dominant group in most populations. There are two main factors; namely increased longevity and declining birth rates. As the number of retired people increases and the number of people of working age decreases, funding of pensions for the elderly becomes increasingly difficult unless those people in work are prepared to accept major tax increases. By the end of the 1990s, observers of population ageing were predicting that many governments would soon be facing potentially massive fiscal crises due to an inability to fund state pensions for elderly people and growing pension deficits associated with pension provision for public sector workers (Jensen et al, 1995). An added factor is that, as people live longer the greater the level of healthcare they require. On average the cost of public pensions and healthcare benefits consumes 12 percent of GDP in developed nation economies. This figure is forecasted to rise to 24 percent of GDP by 2040 (Meeks et al, 1999).

More problems

In the early years of the 21st century the price of oil began to rise dramatically, mainly fuelled by increasing demand from the Chinese and Indian economies (Rees, 2008). Oil is a non-renewable resource and, as the number of cars on the world's roads continues to rise, an increasing demand for petrol can only lead to higher oil prices (Moeller, 2008). This trend will trigger an upward climb in the price of other non-renewable energy resources such as coal and gas. Despite the potential offset available from natural gas that is being produced by fracking technology in countries such as the USA, Hartley et al (2008) forecasted that, with demand for hydrocarbon fuels outstripping supply on a global scale, the world will continue to face an upward pressure on all forms of energy for the foreseeable future.

The world is unlikely to run out of oil any time soon, but interaction between supply and demand will result in organisations facing rising costs for carbon-based fuels over the balance of the 21st century. This will impact other economic activities from the extraction of raw materials and manufacturing of products to production of services and the distribution of goods.

Hence organisations, especially those engaged in high energy usage sectors such as the airline industry, will need to accept rising energy costs as a meta-event, which will have a fundamental influence over future strategies.

The other major natural resource problem is that of increasing shortages in the world's food supplies. The underlying cause is that demand is rising at a rate greater than can be sustained from available supplies (Rosen and Shapouri, 2009). The United Nations have estimated that approximately 40 percent of the world's nations are either at risk from serious food shortages or are already facing starvation. Continued world population growth is an important factor influencing food shortages, but increasing shortages reflect an interaction between various determinants. One factor is a change in eating habits, with rising per capita incomes causing meat to become a more important component of diets in countries such as India, China and Brazil. This switch in eating habits has led to more of the world's cereal output being directed towards the less productive agricultural process of feeding crops to cattle, with 7 kg of grain being required to produce 1 kg of beef. Additional pressure on grain prices has occurred as countries such as the USA have diverted their corn harvest to the highly inefficient process of converting crops into bio-fuels. These are two major factors contributing to 'world food inflation'. Further aggravation in food price volatility has been caused by investors' market speculation (Hojjat, 2009).

Western consumers face further increases in the cost of living following governments' decisions in relation to global warming. An increase in the level of greenhouse gases in the earth's atmosphere, caused by burning fossil fuels, was widely ignored until the publication of reports from the United Nations Intergovernmental Panel on Climate Change (IPCC). Obtaining acceptance of global warming initially proved extremely difficult. Eventually, under the auspices of the United Nations, in 1997 the world leaders met in Kyoto, Japan to discuss proposals about limiting carbon dioxide emissions (Cirman et al, 2009). A significant problem in responding to global warming is the ongoing difficulty of predicting the potential environmental and economic impact. Nevertheless it seems probable that one outcome will be drought in some areas of the world, which will further food price volatility (Kharas, 2011). This effect is already apparent, for example, in the case of the USA, where prolonged drought in 2012 led to an estimated 50 percent reduction in corn and wheat production.

The Kyoto Protocol proposed mechanisms for nations to reduce greenhouse gas emissions, but there have been significant differences in various nations' response to the Protocol. The Bush administration in the USA claimed the Protocol would cost $400 billion in industrial output and a loss of 4.9 million US jobs. Attempts to persuade more Western nations to accept these higher costs were frustrated at the Copenhagen Climate Change summit in 2009 when the developing nations refused to commit to an agreement to meet specific global emissions targets by 2025.

Some governments, which have agreed to meet the Kyoto emission targets, were hoping to exploit nuclear power as a replacement for hydrocarbon fuels. However the problems facing Japan following the Fukushima nuclear plant disaster raised concerns about this solution. Germany decided to close down all of its nuclear energy plants and other EU countries are reconsidering their policies in relation to reliance upon nuclear power. This will mean governments resorting to reliance upon more expensive low emissions solutions such as renewable energy technologies.

In Europe, where acceptance of the need for a reduction in greenhouse gas emissions is high, the economic downturn has caused some companies to express concern about being able to compete with firms in China as they continue to bring new coal-powered generation capacity online. Growing resistance to government policy is illustrated by industry's vocal response to the UK government's 2011 announcement to reduce greenhouse emissions by 60 percent by 2030 (Pickard and Stacey, 2011). Although such criticisms have caused Western governments to reconsider their plans for sustainable energy, any shift away from oil can be expected to further increase consumers' and industry's energy costs.

The storm breaks

By the early 21st century most politicians were aware that factors such as population ageing and healthcare costs meant that governments were facing increasingly unmanageable public sector deficits, yet public sector spending showed no sign of abating (Jackson, 2009). In the UK, government expenditure as a percentage of GDP rose from 33.5 percent in 1965 to 49.7 percent in 2004—and from 33.1 percent to 46.8 percent in the EU and 25.6 percent to 31.3 percent in the USA.

By the late 1980s bankers realised that the declining economic growth in Western democracies would adversely impact the financial services sector. They lobbied politicians to remove the regulations created to avoid a repetition of Great Depression bank collapses by promising this would increase GDP and generate higher corporate taxes. Politicians were persuaded that Central Banks and other regulatory agencies now had sufficient expertise to provide early warnings of potential problems or misbehaviour within the financial services sector to prevent another banking crisis (Kaufman and Wallinson, 2001).

Economists such as Eisner (1996) and Eisenbeis (1997) warned that deregulation, accompanied by low interest rates, ran the risk of fuelling excessive consumer borrowing and a major house price bubble. Rising GDP in the early 21st century was achieved by increased consumer spending, fuelled by low interest rates and higher borrowing by property owners' whose perceptions of higher personal net worth were based upon the increasing value of their houses (Connelly, 2008). What governments did not appear to understand is that, when the world is awash with money based upon uncontrolled lending, monetary policy is unlikely to have a significant impact on any attempts to sustain real economic growth (Nesvetailova, 2005).

These criticisms were very similar to the Austrian business cycle theory (ABCT), developed by Hayek over 60 years earlier and contained within his text *The Road to Serfdom*. Hayek contended that economic cycles occur because productive resources have been incorrectly allocated. Furthermore, he posited that economic bubbles, such as that which occurred in the US housing market, are created through a credit expansion based upon excessive lending and low interest rates. These policies generate unsustainable investments and the inevitable outcome when the bubble bursts is a major economic downturn. Hayek (1991) subsequently extended his observations to propose that excessive public expenditure and taxation by increasing the ratio of spending to saving may cause a prolonged depression and that the duration of that depression should not be counteracted by any inflation of the money supply. Furthermore monetary policy should not be used as an attempt to reduce unemployment in the short run because this distorts the allocation of available resources and will eventually lead to unemployment at a level greater than existed during a downturn (Cochran, 2011).

Bankers in the USA were keen to enter the sub-prime mortgage market, offering low interest loans and reducing the level of security needed to acquire a loan. Poorer people, deluded by bankers' assurances of no risk because the price of property would continue to rise, came to believe they could finally afford to own their own houses. The banks' huge success expanded their balance sheet liabilities, which restricted further lending. The solution was to bundle together these loans in a process known as 'securitisation' and to sell these as collateralised debt obligations (or CDOs) to investors, other financial institutions and pension funds. When the scale of sub-prime mortgage mis-selling emerged many CDOs were recognised as toxic debt. The US banking industry went into crisis mode, some companies such as Lehman Brothers collapsed and others such as Washington Savings & Loan were taken over by other banks. To avert a meltdown, the US government approved a huge bailout fund and implemented a federal lending programme to sustain liquidity in the financial markets.

In the UK, concern over toxic debt led banks to reduce inter-bank lending, which in turn caused short-term interest rates to rise. Banks such as Northern Rock could no longer borrow the funds necessary to service outstanding loans. To avoid a 'bank run' the UK government took over some banks and assisted other banks to buy smaller financially distressed operations. The Bank of England initiated a programme of Quantitative Easing, making billions of pounds available to the financial markets to sustain financial liquidity in an attempt to achieve economic stability.

Within mainland Europe the toxic debts held within the banking sector were not considered to be big enough to require the European Central

Bank (ECB) to initiate actions such as Quantitative Easing. Unfortunately the EU Commissioners had not ensured members of the Euro zone were keeping public spending under control and, as a consequence, when the world economy entered recession, it became apparent that in some European countries governments had only sustained recent economic growth through increased borrowing on the short-term money markets. As the world economy entered recession, it became apparent that in some countries governments had achieved economic growth through increased borrowing on the short-term money markets. By mid-2010 it was clear that Ireland and Greece were in deep trouble and problems of a similar nature were emerging in other Mediterranean countries such as Italy, Spain and Portugal. Initially it was hoped that the International Monetary Fund (IMF) could make sufficient funds available to resolve these sovereign debt crises. However it soon became apparent that the scale of the sovereign debt was beyond the capacity of the IMF and the conventional view that default on government bonds was a problem that only occurred in places such as South America disappeared forever (Bauer et al, 2003).

Euro member nations such as Germany and France were forced to support the European Central Bank (ECB) in the creation of a £1 trillion bailout fund. In order for any country to qualify for rescue funds their government is required to implement massive reduction in public sector spending through the introduction of austerity measures, such as raising taxes, whilst concurrently cutting public sector pay and pensions by up to 30 percent. The response of the electorate in the more stable European economies has been fiercely critical of what they see as their taxes being used to save other, more profligate spenders elsewhere in the EU.

By 2010 it was apparent that the future existence of the Euro was under threat, but politicians remained unable to reach agreement on what should be done. Nations such as Germany resisted the idea that their more successful economies should provide the rescue funds for their poorer neighbours elsewhere in Europe. Governments have fallen and political leadership changed in a number of countries as electorates have sought to punish incumbents and elect new politicians promising to solve the growing economic crisis. By late 2012, the most likely rescue plan is for the European Central Bank, headed by Mario Draghi, to initiate a massive purchase of government bonds with the aim of stabilising borrowing costs for countries such as Spain and Italy. In order for the German government, under Angela Merkel, to support this strategy, countries that purchase their bonds will be required to permit the EU Commissioners to have greater control over the management of their respective economies.

Experts and academics are split over the ECB solution. Some support the view that it will prove effective, others believe that the nature of the Euro zone will undergo fundamental change, whilst others predict the eventual

10 *Entrepreneurship and Innovation During Austerity*

collapse of the currency. Whichever outcomes occurs, there will be an extended period of austerity across most of Europe as governments seek to lower public sector debt through reduced public sector spending and increased taxation.

By remaining outside the Euro zone, the UK was able to avoid significant contributions to the ECB bailout fund. However the Conservative–Liberal coalition, formed in 2010, has been forced to initiate actions to reduce the public deficit through a mixture of public sector spending cuts and raising taxes. Some developed nations, such as Australia, Canada and New Zealand, have to a large degree been unaffected by the banking and sovereign debt crises. Nevertheless the outlook for most developed nation economies is an extended period of austerity involving major reductions in public sector spending, a massive restructuring of public services and many public sector jobs disappearing forever.

Beware Greeks ...

Case Aims: To illustrate some of the problems associated with trying to reduce public sector spending in a Western democracy.

The track record of Greece, in terms of abiding by EU legislation since joining the Common Market, should have provided an early warning of the country's questionable approach to managing public sector finances. Despite this fact European politicians permitted Greece to join the Euro even though the country's public sector expenditure as a percentage of GDP was well in excess of the requirements laid down for membership of the common currency. Hence it came as little surprise that when the sovereign debt crisis emerged in Europe, Greece was a leading participant.

To rescue Greece the ECB and the IMF specified strict requirements concerning actions to reduce the country's public sector deficit. The Greek government agreed to cut 150,000 public sector jobs and achieve annual savings of £700 billion. As demonstrated by the following outcomes, implementation of agreed savings have not occurred at the pace wished for by EU Commissioners (Hope, 2011):

(1) Creation of a unified civil service which was scheduled to commence in July 2011 but was then postponed until October.
(2) Liberalisation of 140 'closed ship' professions and occupations that reduced competition within certain sectors for which legislation was passed but was then not implemented due to objections by groups such as lawyers, taxi drivers and pharmacists.
(3) Privatisation of a number of public sector organisations delayed because of political concerns that the weak global stock market would lead to assets being sold well below their real value.

(4) Modernisation of the tax system involving in part the closure of 200 regional tax offices in 2010. By mid-2011 no offices had been closed due to objections from public sector unions causing the government to decide to delay any action until at least 2012.
(5) Mergers and closures of 150 entities intended to be completed in 2010 and staff being made redundant. Instead staff was transferred to a 'strategic reserve' on 60 percent of their previous salary.

Greece has received significant assistance from the ECB and the IMF but, as demonstrated by these institutions breaking off negotiations in Athens in the summers of 2011 and 2012, these organisations have doubts about the willingness of the Greek government to actually fulfil promises made to reduce the country's public sector expenditure. The Greek electorate's response to a worsening economic climate was to elect a new government, which promised to fulfil undertakings made to the ECB but to reduce the severity of the cuts demanded by the EU. By mid-2012 the new government was achieving no more success in reducing public sector spending than their predecessors. The IMF, European Commission and ECB 'troika' demanded that Greece front-load austerity measures for 2013, because the country will miss the agreed deficit target for 2013. The Greek government had forecasted a fall in GDP of 3.8 percent, but the troika concluded that the fall in GDP is more likely to be in the region of 6 percent. Should this be the case then Greece will yet again be unable to achieve the restructuring of the country's economy that is being demanded (Münchau, 2012). Hence unless countries such as Germany are prepared to provide some form of additional external financial assistance, there remains a high probability of Greece being forced out of the Euro or, even more worryingly, the scale of public dissatisfaction with current events could cause the country to spiral into major social unrest.

Austerity

There are ongoing debates among economists about the causes of the world economy downturn, the benefits of reduced versus increased government spending and the likely future economic outlook. There is reasonable consensus that, in order for current massive consumer and government debts to be repaid, total spending will need to be reduced for a significant period of time. This cutback in spending will inevitably mean that economic growth in some Western economies will probably not re-appear for at least 10 years.

Repayment of excessive borrowing by consumers and governments is only one of the variables that will cause reduced per capita disposable income over the next decade. This is because the following additional factors will further decrease average disposable incomes and lead to a new age of prolonged austerity:

(1) The need to pay higher personal taxes to fund government expenditure on healthcare and pensions.
(2) The need to make higher employee pension contributions because people will live longer and state pensions will fall as governments seek to reduce public sector spending.
(3) Where the welfare state funds healthcare, the need for consumers to make greater contributions to items such as prescription drugs and medical treatment.
(4) Where there is a mixed private and public healthcare system, the need to make higher private pension payments and greater contributions to the cost of treatment as governments reduce the scale of their funding of healthcare services.
(5) The need to make greater contributions to the cost of education as governments transfer more of the costs of education provision to students and their parents.
(6) The rising cost of living due to increasing energy and food prices.
(7) The rise in transportation costs as providers recover higher energy costs by raising the prices.
(8) The decline in the scale of benefits paid to the unemployed or socially disadvantaged as governments seek to reduce public sector deficits.
(9) More people spending longer periods without work, facing a reduction in total income whilst receiving unemployment benefits.
(10) Rising job losses in Western companies as sales decline following foreign firms' entry into the domestic markets of Western producers offering lower priced goods.
(11) Falling corporation tax revenues as domestic firms' profitability declines due to competition from the emerging economies.

Worrying news

Case Aims: To illustrate how the current economic conditions in the UK and USA are influencing consumer behaviour.

UK households' disposable income fell for the second consecutive quarter in the first three months of 2011. During the same period the household savings ratio fell to 4.6 percent, down from 5.1 percent in the previous quarter. This trend reflects consumers' increasing reliance upon their savings to maintain lifestyles. Some economists have predicted that this drop in disposable income and lower savings levels will heighten consumer concerns about the future, leading to further fall in consumer confidence. This outcome will be reflected in consumers' unwillingness to increase their spending even when their disposable income may actually begin to improve (Thomson and Brittain, 2011).

UK families are expected to spend slightly more in 2015 than they did before the financial crisis hit in 2008, reflecting the impact of inflation, tax rises and slow wage growth. The *Financial Times* forecasted that the UK economy is set to experience the slowest recovery in consumer spending for any post-recession period since 1830 (Pimlott, 2011).

In the USA, between May 2009 and May 2011 consumer spending rose by 16 percent among Americans earning more than $90,000 a year, but among all other Americans spending was flat. Any economic recovery during this period appears to have been driven by the spending of affluent consumers. The middle classes, who are the core customers of the nation's products and services, have lost 23 percent of their wealth since the recession began in 2008—versus 12 percent among the upper classes. In reality America's middle classes have faced difficulties for an entire decade, with median incomes declining from 1999 to 2009. This trend was masked by rising house prices, which allowed consumers to enjoy an illusory increase in their standard of living whilst earned incomes were flat or falling. Since 2007 the number of professional, managerial and highly skilled technical jobs has remained relatively unchanged, but over 10 percent of white-collar jobs in sales, administrative support and non-managerial office work have disappeared along with over 15 percent of blue-collar jobs (Peck, 2011). The relative decline in the number of middle skill jobs and slower rises in college completions in the USA means a larger number of workers will be forced to accept low skill service sector jobs, placing further downward pressure on average wages.

Prospects for corporate growth

The success of many Western firms has been based upon decades of growth in the size and spending power of the middle classes in domestic markets. The reduced spending power of this group in the new age of austerity can be expected to dampen sales very significantly for at least a decade. The only source of ongoing growth for these companies will be from sales to the middle classes in emerging economies such as Brazil, India, China and Russia (the BRICs).

Some Western firms have announced a high degree of confidence that long-term growth can be sustained by expanding their operations in emerging economies with specific emphasis on China. This is because China is expected to overtake the USA to become the largest economy in the world. Such optimism may reflect a somewhat limited understanding about how world history has caused China to be xenophobic, an attitude that has evolved over the centuries because the country has suffered very severely at the hands of numerous foreign invaders (Fenby, 2008). China's political leadership warmly welcomes foreign firms where their presence is seen as effective in attracting capital, bringing the latest industrial technology or providing a source of new job creation. However at the juncture where the Chinese government perceive domestic firms can fulfil these roles, foreign corporations usually find remaining in the country becomes increasingly difficult (Chaston, 2009).

A validated example of what may occur in China is provided by events in Russia following perestroika. The Western oil companies were welcomed with open arms to modernise existing fields and to open new ones. Once the Russian government felt their own producers had acquired sufficient expertise, the Western oil companies were confronted with problems, such as exploration licences being revoked and claims that they were avoiding corporation tax (Bucknall, 1997). Ever since, oil companies, and firms from other sectors of industry, have also faced the problem of Russia being a difficult country to do business in. This is due to factors such as the high levels of corruption, the significant control that organised crime has over areas of the country's economy, the poorly developed legal system and a highly autocratic, centralised government that is willing to change legislation whenever this will frustrate the activities of foreign companies (Schlevogt, 2000).

Many business leaders are unwilling to openly express their increasing concerns about the long-term opportunities available to companies operating in China. One of the first sources of adverse comment the CEO of the American GE Corporation, Jeff Immelt, who was quoted as stating: 'It's getting harder for foreign companies to do business there. I am not sure that in the end they want any of us to win or to be successful' (Evans-Pritchard, 2010, p. B2). This view has been echoed by two German corporations, Siemens and BASF, who both feel that 'the playing field in China is increasingly tilted against foreigners'. The European Chamber of Commerce has also concluded that foreign companies are being actively discouraged from investing in China. Traditionally the Japanese have always avoided commenting upon adverse trading conditions elsewhere in the world, but even they have recently begun to express concerns about the risks of making further investments in China (Monaghan, 2010).

Bretnotz and Murphree (2011) undertook an extensive examination of the attitudes and values overseas firms can expect to encounter inside China. They noted that, although the Chinese have become more supportive of greater economic freedom, the government still retains control over the behaviour of the population and the activities of business enterprises.

These authors' analysis revealed that a significant degree of commercial uncertainty exists because the Chinese government is willing to make sudden major policy changes should these benefit domestic organisations or be advantageous in frustrating further expansion by overseas firms. Where a sector of industry is deemed as critical, in terms of supplying domestic markets or expanding into overseas markets, the Chinese government has developed a number of mechanisms to influence outcomes. These include only granting operating licences to certain companies, appointing specific firms as approved government suppliers and refusing to approve applications from foreign firms to open new plants in China.

Another adverse influence is that China's banks remain under state control and hence the government can enforce rapid changes in consumer, industrial and financial markets by requiring the banks to respond immediately to new government directives. In 2011, for example, when the government became concerned over rising house prices they instructed the banks to immediately revise the terms and conditions under which new mortgages could be granted. Although stock markets exist in China, which in theory can provide access to capital, government regulations severely limit the ability of firms to raise capital or engage in merger and acquisitions (M&As).

Design for survival

Some Western firms will continue to enjoy high sales revenue inside China. Others may find China will provide useful source of short-term incremental revenue. Nevertheless the vast majority of Western firms currently succeeding in China, and other BRIC markets, should recognise that their long-term survival is critically reliant upon sustaining sales in their domestic market and in developed economy markets such as Europe and the USA. The problem is that these latter markets consist of customers facing a prolonged period of economic austerity; thereby requiring firms to develop new business strategies capable of responding to a market where customer spending is depressed by lower per capita incomes.

Customers facing austerity become price sensitive and one solution is for firms to accept lower profit margins and generate sales through the use of price cuts and sales promotions. There are problems associated with this solution. Firstly such actions can risk permanently damaging the financial viability of the business as profit margins continue to shrink (Bacot et al. 1992). In some cases this can result in firms exhausting their financial reserves and being unable to service high debt levels or attract new investors. When this occurs the usual outcome is bankruptcy. Secondly, as the BRIC nations expand their international marketing activities by exploiting their lower operating costs as the basis for entering Western markets, domestic firms will be unable to match the lower prices being offered by these market entrants (Small, 1998).

Western firms can learn from Joseph Schumpeter's (1934) conclusion that firms that survived the Great Depression of the 1930s were those which focused on entrepreneurship as the basis for developing innovative new products, entered new markets and, in some cases, created entirely new industries. The American academic Peter Drucker (1985) shared Schumpeter's view about the importance of entrepreneurship and innovation to survive in volatile and rapidly changing markets. Subsequently evidence of the benefits of innovation in the face of increasing price competition has been identified by Gilbert's (1990) analysis of US firms during the 1980s recession. He concluded that management had focused too heavily on sustaining short-term performance whilst cutting back their investment in innovation. Similarly Ghemwat's (1993) study of the US semi-conductor industry during the 1970s recession indicated that to remain successful firms must retain a long-term perspective on how to exploit innovation to sustain competitive advantage.

Key to the Industrial Revolution was the exploitation of technological advances to develop new products and solve complex technical problems by individuals such as 19th-century UK engineer Isambard Brunel and American inventor Thomas Edison. Since this time the Western democracies have retained their leadership in global markets by the exploitation of new knowledge to sustain innovation, in some cases developing products that provided the basis for entirely new industries. Recent examples include Google's exploitation of search engine technology to exploit the Internet and Apple's string of successes with products such as the iPod, iPhone and iPad.

The Western democracies still retain a significant dominance in leading edge science and technology. Until recently a substantial proportion of innovations in areas such as IT, telecoms and the Internet have tended to focus on products or services which appeal to consumers enjoying steadily rising per capita incomes. The arrival of the new austerity changes in most market environments will mean consumers find ways of sustaining their lifestyles in the face of flat or declining spending power. Under these circumstances the new focus of entrepreneurial activity possibly needs to be redirected to developing new products or services which permit consumers to achieve ongoing lifestyle satisfaction despite living in a world where, for the foreseeable future, disposable income will continue to decline. Hence the purpose of the balance of this text is to examine appropriate strategies for Western firms to ensure their survival in developed nation markets by redirecting their innovative activities towards the very different market conditions that are associated with this new age of austerity.

References

Babcock, L., Engberg, J. and Glazer, A. (1997), Wages and employment in public-sector unions, *Economic Inquiry*, Vol. 35, No. 3, pp. 532–543.

Bacot, M.L., Hartman, S.J. and Lundberg, O.H. (1992), Adaptive strategies and survival in an environment dominated by economic decline, *Journal of Applied Business Research*, Vol. 9. No. 1, pp. 34–44.

Bas, D.S. (2011), Hayek's critique of the General Theory: A new view of the debate between Hayek and Keynes, *Quarterly Journal of Austrian Economics*, Vol. 14, No. 3, pp. 288–310.

Bauer, C., Herz, B. and Karb, V. (2003), The other twins: currency and debt crises, *Jahrbuch für Wirtschaftswissenschaften*, Vol. 54, No. 3, pp. 248–268.

Bretnotz, D. and Murphree, N. (2011), *Run of the Red Queen*, Yale University Press, New Haven.

Brown-Collier, E. K. and Collier, B. E. (1995), What Keynes really said about deficit spending, *Journal of Post Keynesian Economics*, Vol. 17, No. 3, pp. 341–356.

Bucknall, K. B. (1997), Why China has done better than Russia since 1989, *International Journal of Social Economics*, Vol. 24, Nos. 7/8/9, pp. 1023–1037.

Chaston, I. (2009), *Entrepreneurship in Small Firms*, Sage, London.

Chaston, I. (2011), *Public Sector Management, Mission Impossible?*, Sage, London.

Cirman, A., Domadenik, P., Koman, M and Redek, T. (2009), The Kyoto Protocol in a global perspective, *Economic and Business Review for Central and South-Eastern Europe*, Vol. 11, No. 1, pp. 29–54.

Cochran, J.P. (2011), Hayek and the 21st century boom-bust and recession recovery, *Quarterly Journal of Austrian Economics*, Vol. 14, No. 3, pp. 263–287.

Connelly, B. (2008), Origins of the credit crisis, *The International Economy*, Vol. 22, No. 4, pp. 44–48.

Drucker, P. (1985), *Innovation and Entrepreneurship*, Butterworth-Heinemann, Oxford.

Eisenbeis, R.A. (1997), Bank deposits and credit as sources of systemic risk, *Economic Review—Federal Reserve Bank of Atlanta*, Vol. 82, No. 3, pp. 4–20.

Eisner, R. (1996), The balanced budget crusade, *Public Interest*, Washington, Winter, No. 122, pp. 85–93.

Evans-Pritchard, A. (2010), Vodafone joins the queue of firms to leave China, *The Daily Telegraph*, London, August 30th, p. B2.

Fenby, J. (2008), *The Penguin Modern History of China: The Fall And Rise Of A Great Power 1850–2008*, Penguin, London.

Ghemawat, P. (1993), The risk of not investing in a recession, *Sloan Management Review*, Vol. 34, No. 2, pp. 51–59.

Gilbert. N, (1990), The time trap: short-term solutions needed for long-term problems, *Management Review*, Vol. 79, No 7, pp. 28–33.

Hartley, P.R., Medlock, K.B. and Rosthal, J.E. (2008), The relationship of natural gas to oil prices, *The Energy Journal*, Vol. 29, No. 3, pp. 47–66.

Hayek, F.A. (1991), *The Fatal Conceit: The Errors of Socialism*, University of Chicago Press, Chicago.

Hojjat, T.A. (2009), Global food crisis—food versus fuels, *Competition Forum*, Indiana, Vol. 7, No. 2, pp. 419–426.

Hope, L. (2011), Venizelos feels the heat of the troika walkout, *Financial Times*, London, September 5th, p. 4.

Kharas, H. (2011), The challenge of high and rising food prices, *Brown Journal of World Affairs*, Vol. 18, No. 1, pp. 97–106.

Jackson, P.M. (2009) The size and scope of the public sector, in Boviard, T. and Loffler, E. (eds), Public Management and Governance, 2nd edn., Routledge, London, pp. 27–40.

Jensen, S., Hougaard, E. and Nieksen, S.B. (1995), Population ageing, public debt and sustainable fiscal spending, *Fiscal Studies*, Vol. 16, No. 2, pp. 1–20.

Johnson, R. (2004), Economic policy implications of world demographic change, *Economic Review Federal Reserve Bank of Kansas City*, Vol. 89, pp. 39–65.

Jordan, J.L., Meltzer, H., Schwartz, A. J. and Sargent, T. J. (1993), Milton, money, and mischief: Symposium and articles in honor of Milton Friedman, *Economic Inquiry*, Vol. 31, No. 2, pp. 197–214.

Kaufman, G.G. and Wallinson, P.J. (2001), The new safety net, *Regulation*, Vol. 24, No. 2, pp. 28–36.

Kim, P.S. and Hong, K.P. (2006), Searching for effective HRM reform strategy in the public sector: a critical review of WPSR 2005 and suggestions, *Public Personnel Management*, Vol. 35, No. 6, pp. 317–327.

Marr, A. (2012), *A History of the World*, Palgrave Macmillan, London.

Meeks, C.B., Nickols, S.Y. and Sweeney, A.L. (1999), Demographic comparisons of ageing in five selected countries, *Journal of Family and Economic Issues*, Vol. 20, No. 3, pp. 223–242.

Moeller, J.O. (2008), *Energy and the Environment, Regional Outlook: Southeast Asia*, Institute of Regional Asian Studies, Singapore, pp. 74–68.

Monaghan, A. (2010) China reassures foreign firms of a level playing field, *The Daily Telegraph*, London, September 8th, p. B8.

Münchau, W. (2012), Relentless austerity will only deepen Greek woes, *The Sunday Times*, London, October 7th, p. 5.

Nesvetailova, A. (2005), United in debt: towards a global crisis of debt-driven finance?, *Science & Society*, Vol. 69, No. 3, pp. 396–419.

Peck, D. (2011), Can the middle classes be saved, *The Atlantic Monthly*, Vol. 308, No. 2, pp. 60–72.

Pickard, J. and Stacey, K. (2011), Huhne pressured to drop climate change targets, *Financial Times*, London, May 10th, p. 4.

Pimlott, S. (2011), Consumer spending recovery set to be the slowest since 1830, *Financial Times*, London, June 1st, 2011, p. 1.

Rees, M. (2008), Just the facts, *The International Economy*, Vol. 22, No. 3, pp. 77–81.

Rosen, S. and Shapouri, S. (2009), Global economic crisis threatens food security in lower income countries, *Amber Waves*, Vol. 7, No. 4, pp. 39–43.

Schlevogt, K.A. (2000), The Russian Federation: time for anti-cyclical investments, *Thunderbird International Review*, Vol. 42, No. 6, pp. 707–718.

Schumpeter, J. (1934), *The Theory of Economic Development*, Harvard University Press, Cambridge, MA.

Small, I. (1998), The cyclicality of mark-ups and profit margins: Some evidence for manufacturing and services, *Bank of England Quarterly Bulletin*, Vol. 38, No. 3, pp. 267–274.

Tanzi, A.V. and Schuknecht, L. (2000), *Public Spending in the 20th Century*, Cambridge University Press, Cambridge.

Thomson, A. and Brittain, A. (2011), Disposable income falls in U.K, *Wall Street Journal*, June 28th, p. 3.

Whalen, C. (2008), The Rubin-Greenspan Legacy, *The International Economy*, Vol. 22, No. 4, pp. 54–58.

2
Customer Values

Cultural values

The behaviour of individuals during the purchase and consumption of goods is strongly influenced by national culture. In terms of understanding why variations in national culture exist, Berges et al (2006) proposed that the following key factors are of significance:

(1) *Religion*: There is a diversity of different religions, with both similar and differing beliefs, across the world. These beliefs impact societal values, often providing the moral norms by which individuals are expected to abide. In certain cases, religious values are such that they also determine the basic economic tenets by which an entire nation's lifestyle is defined.
(2) *Kinship*: These are the values which exist inside the family unit. In some countries the breadth of kinship encompasses an extended group of relatives and spans a number of generations. Kinship is often very strong in Asian countries, which results in the primary loyalty of the individual being to the family and can influence people's behaviour inside organisations. This can be contrasted with countries such as the USA where kinship vales are restricted to family and rarely determine values within the work environment.
(3) *Education*: As individuals participate in education, they are exposed to new knowledge and ideas which may influence their attitudes, possibly leading to a values shift. Some new concepts may contradict existing values that have come from the family or religious beliefs.
(4) *Economics*: The shift from an agrarian to an industrial society causes people to move from rural to urban areas. Where the move leads to an increased standard of living this can cause new values to emerge, for example, materialism may replace community orientated values typically more associated with agricultural societies.

(5) *Politics*: Political systems reflect values and viewpoints concerning aspects of social and national policy. In highly autocratic societies governments usually maintain very tight control over the beliefs and values of entire societies. In democratic systems, differences in viewpoints are permitted and elections can lead to change in a country's political leadership.
(6) *Colonisation*: Groups or entire nations have sought to gain ownership over the resources of other countries. The entrants into a new country often attempt to implant aspects of their home country's culture as part of the colonisation process. Eventually introduced values may merge with existing values within the colonised country; thereby leading to the emergence of a hybrid national culture.
(7) *Immigration*: People are continually engaged in crossing from one nation to another in search of a better life or to escape oppression. Over time, aspects of the immigrants' culture are absorbed into the values of their new location and older values may disappear within one or two generations. This process leads to the emergence of multi-cultural societies.

The Protestant ethic

Case Aims: To present the theories which provide a framework for examining the relationship between religious values and capitalism.

One of the most influential writers on the relationship between culture, religion and business was the German sociologist Carl Weber (Ryman and Turner, 2007). His first contribution, written over 100 years ago, was the *The Protestant Ethic and the Spirit of Capitalism*. This text examined how the Protestant faith influenced the behaviour of European societies and the onset of the Industrial Revolution. Weber's proposed ideas subsequently became more widely known under the generic label of the 'Protestant Work Ethic' (PWE).

Weber posited that the catalyst of modern capitalism, which emerged in Western society in the 16th and 17th centuries, was an entrepreneurial approach to work and the generation of profit. He questioned beliefs embedded in traditionalism which reflect an attitude of work being a reality of life and profit being seen as anti-ethical. The impetus for a societal-wide attitude shift was the Protestant Reformation which led to new values that played a key role in providing the basis for society's acceptance of the benefits of capitalism, encouraged people to work hard, save what they earned and reinvest the profit as a way of fulfilling the purpose of serving and glorifying God. Personal economic success was seen as an indication of God's favour and Protestant congregations were exhorted to be diligent in every aspect of their economic endeavours. According to Weber, economic success coupled with the advent of democracy linked to the rise of science and industrialisation caused these values to be accepted across the Western democracies. Eventually these new values replaced the more fatalistic beliefs prevalent in pre-industrial societies.

Subsequently Weber extended his analysis to other religions to assess whether their values had any influence over which nations were willing to accept industrialisation and capitalism. In his analysis of China, Weber found some compatibility between ascetic Protestantism and Confucianism: both encourage a sober lifestyle, self-control and the accumulation of wealth. However, the key difference he identified was that Confucianism accommodates existing social structures in ways that ascetic Protestantism does not. Confucianism is the religious ethic of the privileged or vested interests, whereby the educated and the patrimonial state supports the dominance of tradition. Within Confucianism, tradition is sacred, there is no acknowledgement of a transcendental God, only a belief in the ancestral spirits to which the individual has an obligation. In Weber's view these differences resulted in significant religious barriers being placed in the way of capitalism in Asia. Technical innovation was discouraged because of the potential to disturb the spirits of the ancestors. Social structures of extended kinship groups protected members against economic competition from adversaries; thereby discouraging repayment of debts, a disciplined approach to work and the adoption of innovative or more efficient work processes. Thus, in Weber's view, Confucianism is firmly rooted in the past and as such does not provide the cultural force sufficient to break free from the traditionalism; thereby blocking the emergence of capitalism.

Subsequent studies by other researchers have not found such a close relationship between the level of the PWE present in societies and prevailing religious values. Nevertheless Weber's writings have provided an effective framework through which to recognise that religious values will influence the behaviour of both individuals and organisations in different nations across the world.

Assessing values

One of the earliest research projects about how cultural values vary by nation was undertaken by Geert Hofstede (1983). He surveyed over 11,000 IBM employees living in 50 different countries and identified the following value dimensions:

(1) *Power Distance*, which is the level of acceptance by a society over the unequal distribution of power within institutions. In the workplace, inequalities in power are exemplified by a hierarchical

superior–subordinate relationship. In countries such as Malaysia and Mexico, employees acknowledge their superior's authority based on their formal position in the organisation's hierarchy. In countries such as Denmark and Israel where people display low power distance, superiors and subordinates usually regard each other as equals.

(2) *Uncertainty Avoidance* refers to the extent to which people in society feel threatened by ambiguous situations. In countries such as Japan and Greece, where there is a high level of uncertainty avoidance, there tends to be strict adherence to prevailing laws, conventions and a strong sense of national pride. Within organisations this value dimension results in formal rules and procedures designed to make employees feel secure and certain within their job role. In contrast in countries such as the UK low uncertainty avoidance means nationalism is less pronounced and questioning of rules is an acceptable behaviour trait. Within organisations located in low uncertainty avoidance nations job roles are less formally structured and relationships between superiors and subordinates are more informal.

(3) *Individualism* refers to the tendency of people to look after themselves and their immediate families, placing these needs ahead of society in general. In countries such as Australia, which prize individualism, personal initiative and personal freedom are considered extremely important. Within countries such as Pakistan, where low individualism prevails, there exist tight social frameworks, an emotional desire to belong and a strong belief in the importance of putting the group before the individual. This latter orientation is known as 'collectivism'. Personal achievement and personal goals, important in high individualism countries, are rarely apparent in collectivist societies.

(4) *Masculinity* refers to the degree to which male values such as assertiveness, materialism and a lack of concern for others prevail. A feminine orientation tends to be reflected by more concern for others, importance placed on relationships and quality of life. In highly masculine societies such as Austria, women are generally expected to stay at home, raise a family and not seek to have a successful career. This can be contrasted with a low masculinity country such as New Zealand, where women have successful careers and there is less evidence of assertiveness being exhibited by their male work colleagues.

(5) *Long-term/Short-term Orientation* is a fifth characteristic which Hofstede identified at a later time in his research activities. This orientation is related to what he perceived as an aspect of Confucian dynamism. Long-term orientation relates to the extent to which individuals accept delayed gratification of material, social and emotional needs. This orientation is common in Asia, inside organisations being reflected in more future orientated managers, emphasis on long-term goals and a willingness to sacrifice short-term success to achieve longer-term benefits. This can be contrast by countries such as the UK and the USA where a short-term orientation tends to dominate.

Cultural capital

Case Aims: To examine the influence of cultural activities on life satisfaction and happiness.

In a post-materialistic society people tend to be orientated toward non-materialistic values such as culture, the arts and the environment. These changing values reflect people's motivations. Once basic needs are met, individuals seek to satisfy higher needs. As a consequence, as incomes rise individuals tend to attach more importance to cultural-artistic activity for satisfying their higher needs. Bourdieu (1973) described this process as placing greater importance on 'cultural capital'.

Cho (2006) concluded that extensive involvement in artistic activities has a positive effect on social cohesion, community empowerment, self-determination and personal identity. Seoyong and Hyesun (2009), in their research on cultural capital in Korea, subdivided the cultural experience into eight activities; namely literature, painting, classical music or opera, performance, theatre, dance performances, film and concerts or entertainment shows. These researchers found a positive relationship between life satisfaction and income, education, health satisfaction, social relations and cultural capital. This result is similar to that of Coffman and Adamek (1999) who concluded that cultural activities enhance the quality of life by way of enlarging social networks.

Changing values

People's experiences will influence their values, attitudes and beliefs, causing them to change over time. Materialist goals are motivated by the need for security but as life becomes financially secure, a materialist emphasis on economic and physical security diminishes. Individuals emphasise post-materialist goals such as freedom, self-expression and the quality of life with some endorsing post-materialist goals consistent with the values of universalism and self-direction. A value shift towards post-materialism reflects the importance given to self-relevant values accompanied by less emphasis on financial security.

A nation's economic development will be reflected in gradual changes in the value priorities of the general public. As employment prospects improve, survival becomes increasingly more certain and reflected in a lessening of emphasis on achieving economic and physical security. People start to be concerned with post-materialist goals (Inglehart, 1999). The advent of the welfare state providing access to higher levels of educational attainment is also associated with accepting post-materialistic values. However this outcome is probably due more to education leading to higher earnings than any change as a result of being exposed to new intellectual concepts.

Materialists tend to feel less financially secure and may give higher priority to income maximisation. These individuals tend to be very concerned about becoming unemployed because this would threaten to lower their standard of living. Post-materialists are usually more financially secure and consider unemployment more as a social issue in terms of exhibiting a sense of solidarity with the less privileged in society. Post-materialists will place increasing importance on non-material quality of life issues. This will be reflected in the demand for higher standards of behaviour from those in positions of responsibility such as politicians and business leaders. Inglehart concluded that, over time, generational replacement will lead to an overall worldwide increase in post-materialist attitudes, because younger birth cohorts reared under conditions of greater economic security will be more likely to exhibit these values than previous generations.

Influenced by Rokeach's (1973) rationale concerning motivational aggregation, Schwartz's (1994) approach to assessing values was based upon a taxonomy defined by motivational domains. By the application of multi-dimensional scaling, Schwartz used motivational domains as the basis for classifying cultures into the following seven types:

(1) *Conservatism*, which emphasises close-knit harmonious relations, maintenance of status-quo and avoidance of actions that disturb traditional order.
(2) *Intellectual Autonomy*, which recognises individuals as autonomous entities who are entitled to pursue their own intellectual interests.
(3) *Affective Autonomy*, which recognises individuals as autonomous entities who are entitled to pursue hedonism and desires.
(4) *Hierarchy*, which emphasises the legitimacy of hierarchical roles and resource allocation.
(5) *Mastery*, which emphasises active personal control over the social environment and individuals' rights to get ahead of other people.
(6) *Egalitarian Commitment*, which emphasises the transcendence of selfless interests.
(7) *Harmony*, which emphasises harmony with nature.

Wilson (2005) concluded that post-materialistic values are consistent with the constructs of wisdom, social justice, freedom and equality; and negatively associated with values that are consistent with materialism such wealth, national security, social order and health. He noted that where individual values have been collapsed into the second-order motivational-level domains, post-materialism emerges, reflecting universalist and self-directed motivations. Materialist values are most apparent among people who are conventional and more interested in self. In contrast universalism occurs as people focus upon the needs of others and are prepared to hold more unconventional views.

Generational values

Case Aims: To illustrate the values that exist between people from different age groups.

Generational history factors such as the economy, scientific progress, politics and technology, or social shocks such as a war, will have significant impact on individuals within a generation (Eisner, 2005). This will be reflected in the different values exhibited by a generation when compared with others from a different age group. These differences can be illustrated by the following generational taxonomy, which exists in relation to the US population (Williams and Page, 2010):

(1) *The Pre-Depression Generation* born before 1930 experienced economic hardships and high unemployment during the Great Depression. Expectations were influenced by World War II and by experiencing subsequent social and technological change. These individuals are conservative, altruistic and tend to be less materialistic as they grow older. Their primary concerns are health, aging, financial and personal security.

(2) *The Depression Generation* born between 1930 and 1945, being small children during the Depression or World War II. Their values include careful consumption, saving, morality and ethics. Conformity, social tranquillity and family are important. In terms of values they are slow to embrace change and many are still in excellent health and quite active.

(3) *The Baby Boomers* were born during 1946–1964 and their parents' improving income and standard of living meant they were indulged as children. Their values include individualisation, self-expression and optimism, defining themselves by their careers and high work ethic commitment. They tend to be self-centred and suspicious of authority.

(4) *Generation X* were born during 1965–1977 becoming adults during a period of difficult economic conditions, which reduced their confidence levels compared to those of their parents. They value family but experience of rising divorce rates has caused many of them to take greater responsibility for raising themselves. Their values are

that of being pessimistic, sceptical, disillusioned and questioning of social conventions. There is a desire to achieve an effective work–life balance.
(5) *Generation Y* were born during 1977–1994, growing up in a time of immense and fast-paced change including more employment opportunities for women and dual-income households becoming standard. The characteristics, lifestyles and attitudes of Generation Y are found among older teenagers and young adults. Their values orientate towards self-reliance, strong sense of independence and autonomy. They have a need for peer acceptance, fitting in socially and being participants in social networks.
(6) *Generation Z* were born after 1994. Their parents married later and are less likely to divorce. This generation has faced global terrorism, 9/11, school violence, economic uncertainty, recession and the mortgage crisis. Generation Z individuals are confident and optimistic, being the 'new conservatives' embracing traditional beliefs, valuing the family unit, self-control and more accepting of responsible. Peer acceptance is very important, with their concept of self being partially determined by their membership in social network(s), increasingly accessed online.

Consumptionism

Most studies of materialism conclude that as incomes and security of employment rise individuals become less interested in acquiring material goods, less concerned about self and more concerned about others. A literal interpretation of this concept of materialism would suggest that inhabitants in countries such as the UK or USA, who have enjoyed decades of increasing wealth, should by now have begun to reduce their levels of consumption. But, on the basis of data on consumer expenditure and ownership of assets such as cars or houses, this clearly has not occurred.

This apparent contradiction has caused some academics to raise the issues of whether (a) the concept of materialism versus post-materialism is the most appropriate model for describing and interpreting consumer behaviour in a developed economy and (b) there are validity problems with the scales utilised to assess materialism (Firat and Schultz, 2001; Yuksul and Mirza, 2010). In examining the relevance of materialism as a basis for determining consumer behaviour, Tanner and Roberts (2000) suggested that the literature tends to present the concept of materialism as a trait that implies avarice, greed, hoarding and selfishness. They concluded that a more neutral interpretation is that of most people preferring to purchase quality, higher priced goods and services, which provide personal comfort and happiness. This can be contrasted with some individuals motivated by a desire to own assets that enhance self image or provide an indication of superior social status. Where ownership is related to a desire to influence the reaction of others by conveying superior status it may be preferable to describe this behaviour trait not as materialism but instead—using the alternative phrase—as 'conspicuous consumption'.

Eastman et al (1999) posited that it is wrong to assume only those with above average incomes buy goods where status is a key factor in the purchase decision. This behaviour is evident across most social classes and is not restricted to extremely affluent people. This observation is supportive of other researchers who have concluded that when consumer income exceeds that of merely being able to purchase goods to support basic day-to-day survival, then purchase behaviour becomes much more complex. Goldsmith et al (2010) concluded from their study of purchase motivations in relation to fashion goods that decisions are rarely related to utility but instead involve the consumer seeking to satisfy personal, experiential needs or are socially motivated reflecting a desire to support the perceptions of the social status of the purchaser. These authors noted that as per capita incomes rise, people become less concerned about the practical aspects of a purchase but instead become more interested in the symbolic value of goods. Ligas (2000) noted that in determining the factors influencing motivation it is necessary to understand the 'consumption goals' underlying the purchase. In some cases these are based upon a required practical benefit, such as the ability of a detergent to wash clothes. In other cases the goal may be that of achieving personal enjoyment, enhancing self image or seeking to influence or persuade others to have a more positive perception about the purchaser.

In view of the complexity of the factors influencing customer behaviour and the apparent contradictions associated with the concept of materialism, companies seeking to determine the influence of the new austerity would be advised to focus on the issue of 'consumptionism'. This perspective is reflected in the research by Eastman and Eastman (2011) who examined consumption and levels of price consciousness and value consciousness in the USA following the onset of the Great Recession in 2007. They concluded that consumers with strong motivation to consume for status reasons are less price and value conscious. This trait can be contrasted with consumers who perceive their lives as being impacted by the economic downturn. Among these individuals there is a significant relationship between the level of consumption and the consumer's view that with declining income and increasing job insecurity it is frivolous to buy status products. Eastman and Eastman also found significant positive relationships between status consumption and level of brand consciousness; namely those who are motivated by status and more brand-name conscious are more likely to perceive that a higher price is indicative of higher quality.

Austerity marketing

The medium term prospect in developed economies is that per capita income after tax and potential purchasing power will decline. Unemployment will remain high because the weak economic conditions will reduce the ability of the private sector to retain staff or create new jobs. In 2008 American and Japanese consumer confidence indexes posted their lowest readings since the surveys began (Izzo, 2008). With confidence remaining low, a fundamental value shift can be expected to emerge with customer attitudes being similar to those last seen in the 1950s (Grossberg, 2009). The American Marketing Association (AMA) has already forecasted that the Western world has moved into a period of 'austerity marketing', which Sullivan (2008, p. 13) defined as 'marketing to consumers who don't want to spend'.

In order to respond to this situation, a deeper level of fundamental strategic thinking is required than merely flooding the market with money-off coupons and announcing price cuts. This is because consumers' new values will be focused upon self and with per capita incomes declining people can be expected to exhibit fewer concerns about the needs of others. More affluent consumers will probably wish to avoid being stigmatised by being better off. This desire will be reflected in a move towards avoiding any flaunting of extravagance and switching spending to items that achieve wealth understatement (Masters, 2008).

Although companies should emphasise greater value in their promotional claims, there is a need to be extremely careful about how to position this proposition. To retain customer loyalty, marketers will need to focus on continuing to fulfil consumers' needs for self-respect, status and self-fulfilment. A values shift towards self will mean that a focus on benevolence, which has been used in recent years for 'green' products, will be less effective. Grossberg (2009) posited that, even in relation to frequently purchased goods, marketers will have to devise promotional programmes that emotionally empower their target customers to indulge parsimoniously. Mui (2008) suggested that, with the Western world entering an age of austerity, the greatest impact will be on the affluent middle-class baby boomers, because this group has faced major losses in the stock market or in the value of their pensions and houses. Under these circumstances their response is likely to be more positive towards promotional messages offering reassurance, better value and self-esteem at affordable prices.

There has been a growth in ethical consumerism, as illustrated by people buying organic food and rejecting the idea of battery-farmed chickens. This trend occurred however during a period of low unemployment, low interest rates and rising house prices. The global recession has been accompanied by increases in the cost of living, tighter lending standards and rising unemployment. These trends have impacted consumer values and, with average consumer per capita income on the decline, growth in the sales of sustainable consumer goods can be expected to slow (PriceWaterhouseCoopers, 2008).

Consumers in many developed nation economies are trading down and staying home, leading to declining retail sales. Carrigan and de Pelsmacker (2009) reported that in France, Belgium and the Netherlands consumers are taking more holidays closer to home. In the UK 65 percent of consumers are buying less luxury food, 53 percent are taking fewer holidays abroad, 51 percent are going out less frequently and 42 percent are planning to spend more time at home. In the USA Barbaro and Dash (2008) concluded that sales have fallen for women's clothing, furniture, luxury goods and airline tickets. Additionally many consumers are switching from brand names to retailers' brands.

The worsening economic picture in Europe and the USA is causing anxiety even among the more affluent consumers and has resulted in changes in purchasing behaviour. The challenge faced by many companies in responding to this change can be illustrated by the problems facing those marketing green products, where increasingly sustaining sales will require making goods available at, what customers perceive to be, affordable prices. Shoppers who still feel they can afford to and want to purchase ethical products, such as organic foods, are being de-motivated by high prices. Research in the UK, for example, suggests that the current price premium for environmentally friendly products is around 45 percent, but consumers are only willing to pay around 20 percent more (PriceWaterhouseCoopers, 2008).

Although consumers still want to buy more ethical or environmentally friendly products, they are influenced by two additional factors; namely confusion and lack of trust. This is because contradictory information about the implications of buying one product over another, such as Café Direct versus Nestlé Partners Blend coffee, is causing even socially aware consumers to feel confused, disempowered and unable to reach a decision (Szmigin et al, 2008). In order to overcome these barriers, Szmigin et al propose that firms need to treat sustainability as an opportunity, rather than a costly add on. This can be achieved by using informational promotional campaigns to build consumer trust; thereby assisting consumers to make purchasing decisions about which products come from sustainable sources.

Austerity and consumer behaviour

Case Aims: To illustrate variation in the nature of consumer behaviour following the onset of the new austerity.

Marketers know that within most markets there are customer groups (or 'segments') who exhibit different preferences which can be attributed to factors such as personal values, prior experience, religion and socio-demographics. Customer motivations and purchasing behaviour within these segments can be expected to change in the face of shifts in prevailing economic conditions. The market research firm Decitica concluded that, following the onset of the new austerity, the adult population in America can now be divided into four groups. The company labelled these 'Steadfast Frugalists' (20 percent of the population), 'Involuntary Penny-Pinchers' (29 percent), 'Pragmatic Spenders' (29 percent) and 'Apathetic Materialists' (22 percent) (Dollivar, 2009).

The Pragmatic Spenders typically enjoy an above average income, and hence Decitica proposed that this group will be extremely attractive to suppliers. Compared to the other segments the Pragmatic Spenders have both the financial capacity and psychological motivation to sustain their pre-recession spending patterns. Decitica noted that within this group 69 percent are highly confident in their ability to control their spending, 73 percent are highly confident about resisting the temptation to spend now and worry later and 59 percent are highly confident of being able to avoid exceeding their spending budget. However the current recession has made the Pragmatic Spenders more sceptical about paying a premium price to purchase brand-name products.

Although most Americas have been forced to reconsider their spending priorities in the new world of austerity, the study suggested that the Steadfast Frugalists positively enjoy having to be more economical. The majority of this group only buy goods when on sale, actively seeking out discounts and using money off coupons. Many in this group reported a high level of satisfaction over engaging in this purchasing behaviour. Hence Steadfast Frugalists can be expected to avoid impulse buying and for major purchases a decision is only made after very detailed consideration of all of the available information.

Involuntary Penny-Pinchers are not pleased about the austerity they have been forced to accept following the downturn in the US economy. These individuals tend to earn a below average income and resent that they are compelled to drastically alter their spending habits and lifestyle. Although they are price conscious, the group find no pleasure in having to carefully compare prices prior to making purchases.

The Apathetic Materialists are a group least concerned about the new austerity, because they have been less affected by the economic downturn. In part this is probably a function of age: many in this group are single, aged between 21 and 29 and were already living on restricted incomes prior to the recession. Their apparent lack of concern is reflected by not purchasing goods on sale or using coupons, low participation in price comparison when shopping and a limited ability to live within their financial means.

Entrepreneurial marketing

The depressed economic conditions expected to prevail in the developed nations for the foreseeable future will result in many consumers being very careful about how they spend their money. Grossberg (2011) recommends that in order to sustain sales companies should adopt 'indulgent parsimony' marketing. This involves embracing a position that appeals to the consumers who want to purchase products which are perceived as frugal but concurrently can be considered as offering both superior quality and value. Recommended positioning messages include: (i) 'attractive pricing/worth my buying', (ii) 'cheap/helps me appear smart and thrifty', (iii) 'good value/I'm worth it' and/or (iv) 'an opportunity/a pleasurable experience'.

The potential problem with Grossberg's ideas is that many firms will utilise price cuts and sales promotions to support their indulgent parsimony positioning. Goodell and Martin (1992) concluded that such thinking will lead to actions that may permanently damage the viability of their businesses. Although price cuts may assist a firm to sustain an acceptable revenue flow during a downturn, the strategy will usually be accompanied by a severe reduction in profit margins. Already, in the UK and USA service sectors, businesses have had to reduce prices since 2009 and have subsequently reported a massive fall in total profits.

Lim et al (2008) concluded that sustaining sales through offering low prices reflects managerial thinking that lacks any real understanding that in many cases more entrepreneurial actions will provide greater protection against eroding company profitability. This viewpoint was endorsed by a survey of over 1,000 CEOs of major businesses conducted by IBM (2008). These individuals considered market conditions to be the worst encountered by their organisations since the 1930s. Their perspective is that survival can only be achieved by increasing expenditure on innovation and ensuring an entrepreneurial orientation is embedded across all of their companies' operations.

It was over 70 years ago that Schumpeter (1934) demonstrated that entrepreneurial behaviour is a more effective response to an economic meta-event than attempting to sustain the sale of current products or services through reliance upon reducing prices. In terms of observing the strategies of many companies since the onset of the current global downturn, organisations appear to have ignored the continued validity of Schumpeter's conclusions. These organisations would also appear to have ignored the evidence from prior post-war recessions indicating that innovation is the most effective strategy for developing a sustainable long-term response to an economic downturn. Companies do need to recognise the reality that customer values in many developed countries are changing in the face of market conditions that (a) young people have never previously experienced and (b) older people last encountered in the late 1940s and1950s. Hence, given the major influence that shifting consumer values will have on purchase behaviour, evidence from both the Great Depression and previous recessions since World War II suggest that, in order to survive, organisations should look towards adopting a more entrepreneurial and innovative marketing orientation.

Brand leader thinking

Case Aims: To illustrate how one major consumer goods company is revising strategic thinking in response to the emergence of the new austerity.

Although radical innovation may offer new long-term opportunities, it would be an error to assume that entrepreneurial thinking cannot also be utilised to enhance the appeal and performance of existing products during the current new austerity. This perspective is supported by Gianni Ciserani, the president of the Western Europe division of Proctor & Gamble (P&G) (Mitchell, 2009). The company has concluded that, in Europe, consumers can be divided into three types, namely:

(1) The experiential shopper who desires and can afford premium performance.
(2) The quality shopper who values brands that can be trusted to deliver on performance claims.
(3) The simplicity shopper who buys only what is needed to fulfil basic needs.
(4) The price sensitive shopper who prioritises purchases on the basis of affordability.

Prior to the global economic downturn, the size of all four segments was approximately the same, implying that half of consumers judged performance as a priority, whilst the other half were more concerned about price. Following the advent of the Great Recession all four segments are placing a greater focus on value. The number of simplicity shoppers has decreased and the total number of consumers for whom price is a priority has risen.

P&G's entrepreneurial response has been a tiered pricing model. This has necessitated the development of new products targeted at the middle-market consumer who is seeking performance and value. The first new product to be launched was Pampers Simply Dry, for consumers seeking a lower price product still capable of delivering the benefit of their child remaining dry overnight. This new product complements the existing premium brands, Pampers Baby Stages and the Pampers Baby Dry. First launched in Germany, the new brand generated incremental sales indicating that new customers were being attracted. This achievement of avoiding brand cannibalisation has prompted P&G to start adding a new value tier to most of their brands across 17 countries.

The risk of increasing the number of price points for a product category is to create greater stocking complexity for retailers and potential confusion in the minds of the consumer. Ciserani believes that, in the past, major companies such as P&G gave insufficient consideration to how innovation can create greater complexity. In his view, the onset of the Great Recession means that companies are being forced to accept that more choice and variety is no longer a viable marketing strategy. Hence firms will reduce their reliance upon developing flankers and product variants to gain more shelf space in the hope that this will lead to higher sales. Ciserani believes the shift towards greater simplicity among consumers means that major brands need to cease product duplication and the introduction of 'me too' propositions. Instead innovation should focus on assisting consumers to survive during what is likely to be a long period of austerity.

Ciserani considers entrepreneurship to be a mandatory requirement for responding to global warming. In his view, this move will require an approach involving complete life-cycle analysis in relation to product utilisation. He quotes the example that in fabric care, 70 percent of carbon emissions are generated by consumers' washing machines, not in the production of the soap powder, product distribution, recycling or packaging. In recognition of this reality P&G's new Ariel Excel Gel is designed to deliver the same performance but in colder water; thereby reducing a washing machine's energy consumption by 50 percent. In addition to developing new products there is a need to develop an 'end-to-end logistics' philosophy in which manufacturers and retailers collaborate to develop solutions which can reduce energy consumption and green house gas emissions.

Similar to P&G, Unilever is also modifying the company's brand strategy to respond to the growing levels of austerity in markets such as Europe. The company is drawing from experience acquired from lessons in Asia to respond to changing consumer shopping habits as the financial crisis impacts living standards in countries such as Greece and Spain. The head of Unilever's European operations commented that 'poverty is returning to Europe' (Chan, 2012). In Spain, the company is now selling Surf detergent in packages for as few as five washes; whilst in Greece, Unilever now offers mashed potatoes and mayonnaise in small packages and has created new low-cost brands for basic goods such as tea and olive oil.

References

Barbaro, M. and Dash, E. (2008), Recession diet just one way to tighten belt, *New York Times*, April 27th.
Berges, I., Dallo, E., DiNozza, A., Lackan, N. and Weller, S.C. (2006), Social support: a cultural model, *Human Organization*, Vol. 65, No. 4, pp. 420–429.
Bourdieu, P. (1973), Cultural reproduction and social reproduction, in Brown, R.K. (ed.), *Knowledge, Education and Cultural Change*, Taylor and Francis, London, pp. 241–258.
Carrigan, M. and de Pelsmacker, P. (2009), Will ethical consumers sustain their values in the global credit crunch?, *International Marketing Review*, Vol. 26, No. 6, pp. 674–687.
Chan, S.P. (2012), Unilever sees 'return to poverty' in Europe, *The Daily Telegraph*, London, October 9th, p. 4.
Cho, K. (2006), Modes of leisure consumption and cultural capital: Bourdieu's cultural theory, *Tourism Science*, Vol. 30, No. 1, pp. 379–401.
Coffman, D.D. and Adamek, M.S. (1999), The contributions of wind band participation to quality of life of senior adults, *Music Therapy Perspectives*, Vol. 17, No. 1, pp. 27–31.
Dollivar, M. (2009), Marketers that fail to take account of 'the diversity of consumers' recession experiences' will fall short, *Adweek*, New York, December 6th, p. 2.
Eastman, J.E. and Eastman, K.L. (2011), Perceptions of status consumption and the economy, *Journal of Business & Economics Research*, Vol. 9, No. 7, pp. 9–19.
Eastman, J.K., Goldsmith, R.E. and Flynn, L.R. (1999), Status consumption in consumer behavior: scale development and validation, *Journal of Marketing Theory and Practice*, Vol. 7, No. 3, pp. 41–52.
Eisner, S.P. (2005), Managing generation Y, S.A.M. *Advanced Management Journal*, Vol. 70, No. 4, pp. 4–16.
Firat, A.F. and Schultz, C.J. (2001), Preliminary metrics for investigating the nature of the postmodern consumer, *Marketing Letters*, Vol. 12, No. 2, pp. 189–2003.
Goldsmith, R.E., Flynn, L.E. and Kim, D. (2010), Status consumption and price sensitivity, *Journal of Marketing Theory and Practice*, Vol. 18, No. 4, pp. 323–338.
Goodell, P.W. and Martin, C.L. (1992), Marketing strategies for recession survival, *The Journal of Business & Industrial Marketing*, Vol. 7, No. 4, pp. 5–17.
Grossberg, K.A. (2009), Marketing in the Great Recession: an executive guide, *Strategy & Leadership*, Vol. 37, No. 3, pp. 4–8.
Grossberg, K.A. (2011), Indulgent parsimony: an enduring marketing approach, *Strategy & Leadership*, Vol. 39, No. 2, pp. 36–42.
Hofstede, G. (1983) The cultural relativity of organizational practices and theories, *Journal of International Business Studies*, Vol. 14, No. 2, pp. 75–88.
IBM (2008), The enterprise of the future, accessed from www.ibm.com/gbs/uk/ceo-study, January 2011.
Inglehart, R. (1999), Global attitude measurement: an assessment of the world values survey postmaterialism scale, *The American Political Science Review*, Vol. 93, No. 3, pp. 665–677.
Izzon, P. (2008), Japanese consumer survey shows slide in confidence, *The Wall Street Journal*, November 13th, p. 10.
Ligas, M. (2000), People, products and pursuits: exploring the relationship between consumer goals and product meanings, *Psychology & Marketing*, Vol. 17, No. 11, pp. 983–1003.

Lim, S., Ribeiro, M. and Lee, S.M. (2008), Factors affecting the performance of entrepreneurial service firms. *Service Industries Journal*, Vol. 28, No. 7, pp. 1003–1013.
Mitchell, A. (2009), The power of simplicity, *International Commerce Review*, December, pp. 116–121.
Mui, Y.Q. (2008), Tapping into shoppers' psyches: battered retailers turn to sentimental sales pitches, *Washington Post*, November 18th, p. 4.
PriceWaterhouseCoopers (2008), Sustainability: Are Consumers Buying It?, PriceWaterhouseCoopers: London.
Rokeach, M. (1973), *The Nature of Human Values*, Free Press, New York.
Schwartz, S.H. (1992), Are there universal aspects in the structure and contents of human values?, *Journal of Social Issues*, Vol. 50, No. 4, pp. 19–46.
Ryman, J.A. and Turner, C.A. (2007), The modern Weberian thesis: a short review of the literature, *Journal of Enterprising Communities*, Vol. 1, No. 2, pp. 175–187.
Schumpeter, J. (1934). *The Theory of Economic Development*, Harvard University Press, Boston, MA.
Schwartz, S.H. (1994), Universals in the content and structure of values: Theoretical advances and empirical tests in 20 countries, *Advances in Experimental Social Psychology*, Vol. 25, pp. 1–65.
Seoyong, K. and Hyesun, K. (2009), Does cultural capital matter?: Cultural divide and quality of life, *Social Indicator Research*, Vol. 93, pp. 295–313.
Sullivan, E.A. (2008), Austerity marketing, *Marketing News*, London, October 15th, pp. 13–14.
Szmigin, I., Carrigan, M. and McEachern, M. (2008), Flexibility, dissonance and the conscious consumer, *Proceedings European Association for Consumer Research Conference*, Milan, pp. 11–14.
Tanner, J. and Roberts, L. (2000), Materialism cometh, *Baylor Business Review*, Vol. 18, No. 2, pp. 8–10.
Williams, K.C. and Page, R.A. (2010), Marketing to the generations, *Journal of Behavioral Studies in Business*, Vol. 3, pp. 1–17.
Wilson, M.S. (2005), A social-value analysis of postmaterialism, *The Journal of Social Psychology*, Vol. 145, No. 2, 209–224.
Yuksul, U. and Mirza, M. (2010), Consumers of the modern world, *Administrative Sciences*, Vol. 29, No. 2, 495–512.

3
Revisiting Management Philosophies

The early years

The theoretical basis underlying mass marketing is that of firms moving from a production to a marketing orientation (Kohli and Jaworski, 1990). Implementation of the earlier mass marketing strategies rested upon the assumption that customer needs were homogeneous and that, by offering standardised products, companies could achieve economies of scale. The high profits generated were then re-invested in heavy promotional spending to build customer loyalty and protect the brand from competition.

Initially in North America, and subsequently in Western Europe, during the 1950s, rising consumer incomes, product affordability through mass production and the introduction of television advertising led to fast moving consumer goods (or 'fmcgs') increasingly being purchased by middle-class families seeking convenience and quality. As sales began to flatten in the late 1950s, major fmcg companies expanded their marketing efforts to both widen their product line offerings to existing customers and seek ways of attracting new consumers, such as those on lower incomes and working women (Hamilton, 2003). By the 1960s there was evidence in some mass market sectors, such as frozen foods, that consumers were beginning to exhibit heterogeneous buying behaviour. This trend reflected an increasing diversity of benefits being sought by consumers due to the influence of factors such as income, age and lifestyle. The major brands recognised they should move away from single standardised product offerings and introduce a wider range of different product propositions. The process known as 'market segmentation' reflected the increasing diversity of customer needs.

By the 1970s, the impact of the OPEC oil crisis and high inflation led to a flattening consumer demand. Marketers seeking to build or protect market share increased advertising spending or expanded their use of sales promotions. In many cases however these tactics had no real impact on market share but instead merely resulted in highly expensive brand wars between companies—for example, those between Pepsi Cola versus Coca

Cola and McDonald's versus Burger King. The primary reason for the failure to increase sales though higher spending was marketers' failure to recognise that customers respond in different ways to promotional activity depending upon their level of product loyalty. Sheth et al (2000) proposed that in order to optimise market performance companies need to undertake a review of the influence of effectiveness and efficiency of their marketing strategies, in relation to customer loyalty, and to focus on marketing campaigns directed towards customers who exhibit high loyalty. They referred to this approach as 'customer centric' marketing.

As summarised in Figure 3.1, where the customer exhibits low product loyalty, increased marketing spend will be of little financial benefit because only among loyal customers is increased marketing activity likely to result in a beneficial outcome. One scenario proposed in Figure 3.1 is where the company has a highly effective marketing operation but marketing efficiency is low. An example of low marketing efficiency is that of branded goods, such as soaps or detergents, where minimal actual performance differences exist between products being offered by different companies. In these cases high spending on activities such as television advertising is necessary in order to create a perception of difference in the customers' minds between a brand and offerings by competition. In contrast where there is a genuine tangible difference between brands, marketing expenditure to sustain customer loyalty will be much lower. This latter scenario usually offers a more profitable organisational opportunity.

By the 1970s, in countries such as the USA and the UK, the retail market had become dominated by the national supermarket chains. With the vast

	Low Efficiency Marketing	High Efficiency Marketing
Low Effectiveness Marketing	Expensive Zero Loyalty Customers	Expensive Limited Loyalty Customers
High Effectiveness Marketing	Expensive Customer Loyalty Retention	High Profitability Customer Loyalty Retention

Figure 3.1 Marketing outcomes

majority of grocery products now being sold through these outlets, these organisations had the power to demand their suppliers offer greater levels of sales promotions in order to retain their in-store presence. In many cases recovery of higher marketing expenditure associated with fulfilling the retailers' demands through a strategy of higher prices proved impossible because the major brands were concurrently facing increasing levels of in-store price competition from the retailers' own private label products. As a consequence most of the major fmcgs ceased to enjoy the high profit margins which had been achievable in the 1950s and 60s (Haines, 2007).

Bank wars

Case Aims: To illustrate how aggressive mass marketing strategies may merely result in reducing profitability or overall organisational viability.

In the late 1980s the financial services sector became more marketing orientated; but instead of learning from the experiences of fmcgs, their tendency was to seek to build market share by outspending competition on advertising and sales promotions. A major problem facing financial services companies such as banks, brokerage firms or insurance companies however is that they are unable to differentiate themselves from competition because they are all essentially offering the same product proposition.

During the 1990s, one country to see intensive brand wars in the financial services sector was the USA, where firms were all fighting for a larger share of the country's $23 trillion in household financial assets (Geer Jr., 1997). The problem facing many of this country's retail bankers was that loyalty to traditional banks had declined as depositors moved from low-return checking accounts and Certificates of Deposit (CDs) to invest in mutual funds and money market accounts. Although the percentage of sales allocated to advertising in the financial services sector remained lower than most of the major branded goods companies, over a five-year period total annual spending in the US retail banking sector increased by 127 percent.

Another problem facing retail banks in most countries is that the average customer is not very keen on moving to another supplier. Banks would prefer to lose customers who are a poor credit risk. The problem is the majority of customers who are persuaded to switch banks tend to be those individuals who are poor credit risks. Hence the typical outcome of bank spending battles is no real change in total customer numbers, a significant reduction in profitability caused by increasing spending on marketing programmes and the acquisition of additional higher credit risk customers.

One of the most successful exploiters of a more aggressive approach to the marketing of financial services in the USA was the brokerage firm

Charles Schwab. As with most cases where early movers' success attracts competition, Schwab soon faced competitors being prepared to ignite a price war. In the case of the brokerage industry this was in the form of competitors such as E-Trade reducing their commission on share trading. In order to avoid responding to this threat by reducing commissions, Schwab made significant investments in the development of a whole range of new funds and advisory services in order to protect the company from price-based competition.

In the UK in the 1990s, the most aggressive institutions in retail banking were endeavouring to build brand share in the retail mortgage market. To achieve this goal, some institutions offered lower interest rates, authorised mortgage loans equal to 120 percent of the value of the property and permitted self-employed people to self-certify their income when seeking a loan. When deterioration in the sub-prime mortgage in the USA triggered a global financial crisis, the most aggressive UK lenders were immediately in trouble. The market leader, Northern Rock had to be taken over by the government in order to avoid a 'bank run'. A number of smaller players only survived by being taken over by larger financial institutions (Chaston, 2010).

Market leadership in the UK mortgage sector had been based upon attracting more consumers to open savings accounts, which in turn permitted the institution to fund more mortgages than the competition. Mr Applegate, then CEO at Northern Rock, had the apparently brilliant idea of borrowing money in the short-term money markets where prevailing interest rates were much lower than rates being paid to savers. These cheaper funds could then be used to offer mortgages at an interest rate lower than the competition. As Northern Rock embarked on a battle for market share Mr Applegate was lionised in the financial press for bringing more aggressive, modern thinking into the conservative world of UK banking (Urry, 2003). All was going to plan until, in 2007, the sub-prime mortgage crisis in the US caused European banks to become wary about lending money to each other. Money became scarce and short-term interest rates rose dramatically, leaving Northern Rock in the position of being unable to pay off loans that were coming due or to raise additional loans to service the rising costs of money market debts which had already been incurred. As word spread about the problem there was a run on Northern Rock as worried UK consumers rushed to remove their savings. The queues that formed outside Northern Rock branches represented the first bank run in Britain since 1866 (Anon, 2007).

The lesson organisations should learn from observing the outcome of the banking wars, and also the numerous brand wars between companies such as Proctor & Gamble versus Unilever, that have occurred over the last 50 years is that the best response to a brand war is usually to avoid getting involved, wait until the dust has settled and then move in and exploit the opportunities created by firms whose financial positions have been weakened. Despite the well established validity of this concept it is often the case that CEOs of large companies perceive a brand war as a personal affront to their reputation, and hence seek to become embroiled in the battle. Such was the case with HBOS in the 1990s, which was created when Royal Bank of Scotland acquired Halifax, one of the UK's largest mortgage lenders. Then HBOS CEO, Mr Andy Hornby, had a reputation for being extremely aggressive in his response to any threat from competitors. Hence when it was understood that Northern Rock was achieving market share growth by borrowing short-term funds via the money markets, Mr Hornby's reaction was to duplicate the model and, as a result, in 2008 faced similar problems raising new loans in the short-term money market to service the bank's large mortgage portfolio. Following a massive collapse in the value of HBOS shares, the only viable solution was for the UK government to assist the bank to be taken over by the more conservatively managed Lloyds-TSB (Anon, 2008).

Another common response of CEOs seeking to become market leaders or retain market leadership when their organisation is no longer capable of growing the existing business, is to embark on a strategy based upon aggressive acquisition of other companies. An example of the hazards of this philosophy is provided by Sir Fred Goodwin during his tenure as the CEO at Royal Bank of Scotland (RBS) (Chaston, 2010). Goodwin joined RBS in 1998 shortly after the bank had purchased National Westminster Bank in 2000. This was a hostile takeover of an English bank twice its size and was an incredibly audacious move. As the new CEO, Goodwin set about integrating the two businesses, reducing costs and then using the enlarged business base to move RBS from being a domestic player into a global one. The bank spent billions buying banks in America and a stake in the Bank of China. By 2004, RBS enjoyed a market value of $70 billion, outpacing JP Morgan Chase, Deutsche Bank, Barclays and UBS. Goodwin was named Global Businessman of the Year by *Forbes* magazine in 2002 and received his knighthood for services to banking in 2004.

In 2007 RBS purchased the Dutch bank ABN Amro. The deal was a milestone in the banking industry, being the world's largest financial services acquisition and the first hostile cross-border takeover of a sizeable mainland European bank (Larsen, 2007). A key driver in terms of being prepared to outbid competitors was Sir Fred Goodwin. The business media hailed him as an outstanding leader who was assisting the UK banking industry to retain the UK's rightful title as a major player in the world financial markets.

In January 2009 shares in RBS collapsed when the UK government was forced to mount a rescue bid after it became apparent that the bank was facing the worst losses in the UK's corporate history (Anon, 2009). The UK government took a 70 percent equity holding in the bank. The

> government's negotiations included the departure of Sir Fred Goodwin on the grounds that a significant proportion of the blame for the bank's failure could be attributed to his decision to embark on an excessively aggressive acquisition programme. Public acrimony was generated when it was revealed that Sir Fred Goodwin, in agreeing to step down as CEO at the age of 50, would receive an annual pension of £693,000.

Relationship marketing

As the largest economy for over 100 years, the USA has understandably been the major source of modern management theories. A long established aspect of American managerial theory is the concept of success being achieved by engaging in adversarial activities by defeating competition and controlling the behaviour of suppliers. The validity of this concept was challenged by the Nordic school of management in their International Marketing and Procurement (IMP) research project. They concluded that success of some firms was influenced by collaboration with other organisations (Gummesson, 1987). Initially these concepts were primarily attributed to industrial markets where buyer–seller interactions were orientated towards joint optimisation of performance and groups of companies created network entities to assist activities such as entering new overseas markets.

Further support for the benefits of non-adversarial relationships emerged when researchers examined managerial practices in consumer service markets. Gronroos (1989) posited that traditional marketing theory focused on short duration actions designed to generate the next purchase. He argued that this approach, subsequently labelled as 'transactional orientation', was inferior to the preferable philosophy of seeking to achieve long-term loyalty based upon the creation of a mutual commitment between the organisation and the customer in which both parties undertook to fulfil clearly defined promises. This new approach subsequently became known as 'relationship marketing'.

Morgan and Hunt (1994, p. 24) proposed that 'relationship marketing refers to all marketing activities directed towards establishing, developing, and maintaining successful relational exchanges'. Support for the concept of relationship marketing was presented by Reichheld and Sasser (1990) who posited that transactional marketing, which merely focuses upon the generation of the next sale, has no regard for the fact that the lifetime value (LTV) of customers is more important than generating the next purchase. In their view LTV requires that firms should focus upon retaining customers through adopting relationship orientated actions in order to build customer loyalty. Gummerson (1994) posited that organisations should be concerned with the management of the relationship-based marketing mix, which he entitled the '30 Rs'.

Hunt et al (2006) suggested that the philosophy of relationship marketing can only be effective where:

(1) The supplier can be trusted to reliably, competently and non-opportunistically provide an appropriate product or service.
(2) The supplier and the customer share the same values.
(3) The customer's search costs are decreased over time.
(4) The customer perceives a decrease in risks by participating in the relationship.
(5) The exchange has the potential for customisation that can result in more effectively satisfying customer needs.

A tendency of some academics upon embracing a new theory is that their enthusiasm causes them to totally reject existing theories. Such has been the case with relationship marketing, where some academics saw the concept as a 'paradigm shift' rendering redundant concepts based upon transactional marketing theory. Such tendencies are extremely dangerous when they mislead the management practitioner (Gummesson, 1994). For example, Reichheld and Sasser had posited that focus upon retaining customers is always beneficial. Subsequent research contradicted this, demonstrating the possibilities of adverse cost–benefits outcomes when forming a relationship with certain customers will deliver satisfaction but erode profit margins.

To dispel incorrect assumptions over the superiority of relationship marketing, Gronroos (1997) suggested that markets and customers exist on a continuum. At one extreme are situations where the only effective strategy is to use a transactional marketing philosophy and to merely focus upon generating the next sale. At the other end of the continuum, where there is evidence of high lifetime customer value, the focus should be on a complete commitment to relationship marketing. Between these two extremes Gronroos suggested that a mix of transactional and relationship marketing activities is more likely to optimise cost–benefit outcomes for both the buyer and the seller.

Whether a firm embraces transactional or relationship marketing makes no real difference to the reality that all products and services will reach the point where sales and profitability decline, leading eventually to the demise of such goods. There are a number of factors influencing this eventual demise. For example, changes in fashion (e.g. the crinoline, the whale bone corset), technological advances (e.g. the slide rule being replaced by the electronic calculator) or changing market structure (e.g. small specialist shops being usurped by large national retail chains). Whatever the reason, firms that ignore such changes, seeking to sustain their existence by marketing the same, unchanging products or services, will themselves eventually cease to exist.

Theodore Levitt (1960) labelled this lack of recognition that markets are changing and the business strategy is not revised as 'marketing myopia'. His suggested that myopia occurs because firms fail to recognise the real nature of the business they are a part of, and hence ignore the emergence of new competitive threats. To avoid myopia and ensure long-term survival usually requires firms to adopt an entrepreneurial orientation and exhibit a strong commitment to engaging innovation. This is because innovative firms tend to be (i) better at responding to changing consumer needs, (ii) more successful, (iii) more profitable and (iv) able to exploit new windows of opportunity to replace existing products or services before these become financially unattractive propositions (Cooper,1994; Cooper and Kleinschmidt, 1995). In some cases innovation can be based upon re-investing in improved products to keep the company ahead of the competition. An example is provided by Microsoft which having become the market leader in PC software has regularly launched upgrades to the company's Office product range.

Re-invention

Case Aims: To illustrate that in certain circumstances company survival will require a fundamental change in strategy and market positioning.

In some cases the scale of the competitive threat requires the organisation to be prepared to 're-invent itself'. Such was the case with IBM, which in the 1990s failed to respond to the growing importance of PCs and a decline in market demand for mainframe systems. The company took the unusual step of recruiting somebody from outside the industry; namely Lou Gerstner. Within a very short period of time he announced that IBM needed to introduce fundamental change and stop thinking it was in the computer hardware business. His repositioned the company as the organisation capable of solving clients' complex data storage, handling and analysis problems.

Another example of re-invention is provided by Apple. Whilst still a leading innovator in the PC industry the company fired the founder Steve Jobs for being too unconventional and undisciplined. By the time he was re-hired, the company was a shadow of its former self. Fortunately Jobs was a visionary who understood that the Internet had changed the world of electronic communications for ever. This understanding led to Apple developing three world beating products; the iPlayer, the iPhone and the iPad, which in 2011 resulted in Apple, on the basis of the quoted share price, briefly becoming the most valuable company in the world.

Entrepreneurship

Individuals such as Steve Jobs are excellent examples of how an entrepreneur can radically alter the business prospects of existing firms and, in some cases, implement ideas that create entirely new sectors of industry. Entrepreneurs such as Jobs tend to see the world through very different eyes than the majority of people. As a consequence they have the rare ability to challenge conventional thinking or business practices, resulting in the creation of very different products, services or internal production processes.

For companies in the developed economies wondering how they can survive in today's crisis filled world, it is worth remembering that the problems which emerged during the first decade of the 21st century are not new. During the Great Depression of the 1930s banks went bust and unemployment rose to almost 25 percent of the total work force in many nations. At that time these events prompted a number of academics to re-examine the validity of capitalism as a viable economic principle. One group, the 'Austrian School of Economics', were profoundly concerned by the future of capitalism in the face of widespread acceptance of pro-socialist models such as Communism in Russia and the emergence of extreme right wing, authoritarian regimes in Western Europe. One of the most influential theorists within the Austrian School was Joseph Schumpeter (1950). He held the view that capitalism can be expected to go through periods of 'creative destruction'. This is caused by organisations becoming too set in their ways and being unable to implement the changes required to sustain their ongoing survival. Schumpeter posited that the trigger for creative destruction and economic change is innovation. This causes existing, non-effective institutional frameworks to be replaced by solutions offering superiority over existing propositions. If one accepts Schumpeter's perspective then clearly the solution for both individual organisations and entire economies in the 21st century must lie with becoming more entrepreneurial.

Post-war supporter of the Austrian School of Economics Israel Kirzner presented a theory of entrepreneurship offering a perspective somewhat different from Schumpeter's theories (Langlois, 2007). Kirzner perceived that the dynamics of competition pushes economies toward equilibrium and that the environment facing the entrepreneur is one of imperfect information about prices and organisational strategies. He argued that the entrepreneur is an individual who purposively changes prices, quantities and other aspects of market systems (Kirzner, 1973). A key attribute is that of 'alertness to disequilibria', which permits the entrepreneur to recognise the profit opportunities that exist due to differences within and between markets. The opportunities available for entrepreneurship are provided by future uncertainties. Kirzner (1992) proposed market environments where individuals face genuine uncertainty and are usually ignorant of many of the factors that can influence economic outcomes. In these circumstances an action which is entrepreneurial will involve individuals overcoming their lack of knowledge by being prepared to act in a way that they hope will have a desirable outcome.

Kirzner's view of the entrepreneur moving an economy towards equilibrium can be contrasted to the Schumpeterian entrepreneur who is presented as source of disruption in relation to existing equilibria. A characteristic of Schumpeter's entrepreneur is sufficient boldness and resolve to be able to overcome social or market resistance when engaged in successful innovation. In contrast Kirzner posited that vision, not self confidence, boldness or courage is the requirement to achieve an entrepreneurial outcome. Attempts by other academics to bridge the apparent gap which exists between Schumpeter and Kirzner's views of entrepreneurship have not proved totally successful. Holcombe (1998) attempted to extend Kirzner's theory into a long-run growth model, which is an outcome important within Schumpterian theory. He concluded that long-term growth occurs due the self-reinforcing process whereby entrepreneurial discoveries stimulate the identification of new opportunities. This is a dynamic process that can be influenced by economic conditions, organisational environments and cultural values. In relation to such observations and the recognition of gaps in entrepreneurial theory, Chaston (2010) drew upon Covin and Slevin's (1988) viewpoint to propose that entrepreneurship exists on a continuum. At one extreme innovation outcomes are of the type supported by Kirzner's perspective and, at the other extreme, innovation is of the disruptive form posited by Schumpeter.

Given the importance that major corporations need to place upon entrepreneurship, it is somewhat disconcerting to find this topic is an area of management theory which is apparently sometimes misunderstood by academics. In recent years, for example, there has been a tendency for the term to be used interchangeably with that of small business management. In reality most small firms are in no way entrepreneurial. This is because their operations merely duplicate the activities of numerous other businesses operating in the same market sector. A more informed view of entrepreneurship is that proposed by those academics who have sought to understand the process through research. They have concluded that there is no relationship between the size of a firm and an ability to be entrepreneurial (Hisrich and Peters, 1992).

Miller (1983) suggested that the entrepreneurial orientation of a firm is demonstrated by the extent to which top managers take risks, favour change and exploit innovation to achieve a competitive advantage. Hills and LaForge (1992) concluded that being a successful entrepreneur requires the presence of certain attributes; namely an ability to exploit innovation and to develop a unique operation that supports business growth. Georgelli et al (2000) proposed that the skills of entrepreneurship are a capacity for changing business processes and the launching of new products or services. Covin and Slevin (1988, p. 219) defined entrepreneurial orientation in terms of the extent to which 'managers are inclined to take business-related risks, favour change and innovation, and compete aggressively with other firms'.

Business recession proofing

Case Aims: To illustrate that even entrepreneurs accept there may be a need to modify organisational strategies in today's troubled economic conditions.

Astute entrepreneurs recognise that ideas which might succeed when economies are growing may need to be replaced with different concepts during a major economic downturn. In their study of entrepreneurs in Silicon Valley, California, Eggers and Kraus (2011) described this behaviour as individuals and firms redirecting their efforts towards developing 'recession-resistant offerings'.

Identified activities include rapidly developing product modifications, which will appeal to recession-resistant target groups who although remaining affluent have become more cautious in their buying behaviour. One simple approach is to increase perceived value by changing the appearance of the product and adding extra applications which actually cost very little to create. Concurrently firms may use de-engineering to develop and launch less expensive products to satisfy those customers who no longer feel able to purchase the latest generation products. Eggers and Kraus concluded that these actions do little to change the core product proposition but instead are designed to ensure the generation of an ongoing revenue stream. In some cases entrepreneurs will decide to postpone the development of new-to-the-world propositions until there is evidence that an economy is beginning to exhibit a certain degree of recovery.

Entrepreneurs who contributed to the study confirmed that customers and intermediaries do become more price sensitive in a recession. Hence, when the producer feels unable to offer a reduced price, customers will need to be offered additional product or service value. Where a change in the price–value relationship has emerged one solution for generating additional sales is the launch of a 'freemium model'. These are models where certain services are offered for free and the customer only has to part with any money when they decide to purchase extra additional 'premium' services.

The high costs of developing products based upon leading edge technology means that during a downturn many Silicon Valley firms implement actions to reduce spending on product development (known as the 'burn rate') in order to minimise overhead costs. As many of these firms have limited financial reserves, cost reduction accompanied by action to generate more revenue from existing products becomes a necessity. One way of reducing costs is to save on salaries by paying employees with equity, accompanied by a shift towards hiring new staff prepared to work for equity or even for free by accepting posts as temporary company interns.

One of the critical factors influencing the success of many Silicon Valley firms over the years is the high level of networking. These networks become even more critical during a recession by permitting firms to collaborate with others to develop, promote and distribute products. Network membership is also useful for obtaining access to funding, employees, new co-founders and external advisors. The conclusion reached by Eggers and Kraus is that the current recession has caused many of Silicon Valley's entrepreneurs to shift from their preferred aggressive growth strategy to a strategy based upon survival in the face of flat or declining market demand and limited internal financial resources. Innovation continues but the planned outcomes tend to be of a more conservative nature. The study demonstrated that entrepreneurs are very aware that, in uncertain times, the preference for relying upon intuitive idea generation must be accompanied by a realistic appreciation of changing customer needs.

Entrepreneurship and planning

Most academic management texts are supportive of the idea that organisations can benefit from engaging in formal strategic planning. A commonly described model provides the basis of three basic questions; namely (i) 'where are we now?', (ii) 'where are we going?' and (iii) 'how are we going to get there?' (Johnson and Scholes, 1999).

Despite the extensive support strategic planning has received in the academic literature, the question still exists whether a strategic plan can have a positive influence on the performance of entrepreneurial firms (Shrader et al, 1989). Some researchers, through observations of the actual behaviour of entrepreneurs, have not been able to reach any definite conclusions on this issue (Carson, 1985). Allison et al (2000) have concluded that few entrepreneurs engage in formalised, long-term planning or use highly structured, logical models in managing their businesses. Instead there is a tendency to act on the basis of instinct, intuition and impulse.

Mintzberg (1979) was a strong supporter of the perspective that successful entrepreneurs often avoid utilisation of detailed business plans during the development of future strategy. Drucker (1994) was another very influential academic who was supportive of Mintzberg's ideas, being somewhat sceptical about the practical benefits of highly structured, linear planning. He concluded that long-range planning in many firms is merely concerned with defining organisational policies and the analysis of existing information. Ducker's view was that planning should involve acquiring new knowledge to develop a better understanding of potential future environmental conditions as the basis for identifying the most appropriate future opportunities available to an organisation.

Where an organisation has adopted an entrepreneurial orientation but prefers to sustain a philosophy of developing a strategic plan to guide future activities, the issue arises of what is the most suitable planning model. Anderson and Atkins (2001) concluded that the classic linear sequential model offers limited benefits to entrepreneurs. This is because the existence of complexity and interaction between the variables often demands simultaneous consideration of all factors when determining future actions. As a consequence entrepreneurs who utilise some form of structured planning process tend to adopt what Chaston (2010) refers to as a 'spider's web' approach. In the age of austerity the three questions which the planning process seeks to answer probably need to be posed somewhat differently. This is because merely extrapolating from the past is no longer a relevant or viable process. Instead, as summarised in Figure 3.2, in adopting the spider's web approach the questions which now need to be asked are:

(1) What is the nature of tomorrow's world?
(2) What gaps exist in relation to tomorrow's market and capability versus the current market and capability?
(3) What viable objective(s) can be specified by the organisation in tomorrow's world?
(4) What strategic response will best support achievement of this objective(s) in tomorrow's world?

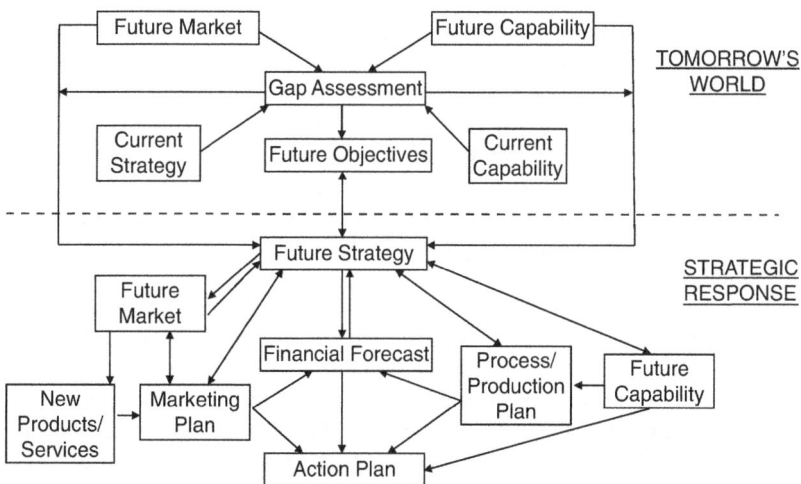

Figure 3.2 Possible components in an entrepreneurial approach to planning for tomorrow's world

Innovation partnerships

Case Aims: To illustrate how a collaborative approach to entrepreneurial activities can assist firms to overcome the problems of sustaining sales in increasingly price sensitive markets.

Long before the latest downturn most companies had accepted that survival in a volatile, rapidly changing world requires attributes to be flexible and proactive. One lesson learnt over the years is that sustaining these attributes demands the establishment of highly effective, efficient supply chains. The operation of such supply chains can only occur if there is mutual trust and commitment between all participants. The advent of an economic downturn may result in firms which have the greatest power within the supply chain reverting to the exhibition of an adversarial attitude involving placing pressure on suppliers to reduce prices or make a financial contribution to some aspect of the company's operations. An example of this behaviour trait is already appearing in some retail supply chains, as exemplified by some major supermarkets having forgotten earlier claims of wanting to work in close collaboration with suppliers and instead using their power to insist that to avoid being de-listed suppliers must agree to price reductions or making some form of financial contribution. In the UK, for example, M&S not only pressured suppliers to offer better terms but also asked suppliers to make a contribution equal to 1.25 percent of their sales to assist the retailer's planned £600 million refurbishment of their major stores (Felsted, 2011). Retailers such as Tesco and Sainsbury may have also requested their suppliers to offer better terms on future sales. There are clearly some adverse long-term implications of such behaviour, because trust or commitment which may have existed in the past can be expected to disappear. This could reduce the efficiency and effectiveness of future supply chain operations. Furthermore forcing suppliers to offer better terms may reduce these organisations' profitability and inevitably result in reducing their abilities to invest in innovation. The ultimate casualty will be retailers, who will have fewer opportunities to attract and sustain customer loyalty through the regular on-shelf provision of new and improved consumer products.

Suppliers understand that a prolonged period of austerity will result in demands from intermediaries to reduce prices and responding to such pressures can, over the long term, severely weaken an organisation's ability to survive. Hence their response should be to examine how innovation involving other upstream and downstream members of their supply chain can permit avoidance of demands to compete on the basis of lower prices.

An example of placing greater reliance on innovation in response to the latest recession is provided by Sonoco, one of the world's largest manufacturers of industrial and consumer packaging based in Hartsville, South Carolina, USA (Slater, 2010). Following the global downturn, consumers switching from branded to private label products caused Sonoco customers in the food and drinks industry to consider switching to more economical packaging designs and move from long-term to short-term procurement contracts. To survive price-based competition and avoid price wars, Sonoco adopted a strategy based upon adding value through innovation involving customer collaboration and enhanced service levels.

An example of collaborative innovation is provided by the company's partnership with a manufacturer of breakfast cereals to develop effective packaging which ensured the product remained fresh for a longer period after being opened by the user. This led to the development of Linearpak containers which have their rounded edges and plastic overcaps. The package clearly stands out in stores and provides consumers with a product which stays fresher longer.

Another area where Sonoco recognised that it was possible to add value to an existing key customer relationship is through more accurate forecasting of demand. Average forecasting error was running at 60–120 percent on a 30-day basis, which had negative implications for inventory levels, meeting delivery dates and sustaining customer service levels. Sonoco discussed the problem with a key customer and they agreed to share data which was confidential and competitively sensitive. Following agreement over process management, data exchange was initiated which greatly reduced forecasting errors and generated operational cost savings for both organisations. This success led to Sonoco implementing their Vendor Managed Inventory (VMI) programme, which was offered to all major customers.

References

Allison, C., Chell, E. and Hayes, J. (2000), Intuition and entrepreneurial behaviour, *European Journal of Work and Organisational Psychology*, Vol. 9, pp. 31–42.

Anderson, A.R. and Atkins, M.H. (2001), Business strategies for entrepreneurial small firms, *Strategic Change*, Vol. 10, No. 6 , pp. 311–324.

Anon (2007), The Bank that failed—Britain's bank run, *The Economist*, London, pp. 13–14.

Anon (2008), Our model was too dependent on wholesale funding, says Hornby, *The Daily Telegraph*, London, September 19th, p. B3.

Anon (2009), Fiscal performance and challenges, *OECD Journal on Budgeting*, Vol. 9, No. 1, pp. 23–43.

Chaston, I. (2010), *Entrepreneurship and Small Firms*, Sage, London.
Cooper, R.G. (1994), Third generation new product processes, *Journal of Product Innovation Management*, Vol. 11, pp. 3–14.
Cooper, R.G. and Kleinschmidt, E.J. (1995), Benchmarking the firm's critical success factors in new product development, *Journal of Product Innovation Management*, Vol. 12, No. 5, pp. 374–391.
Covin, J.G. and Slevin, D.P. (1988), The influence of organizational structure on the utility of an entrepreneurial top management style, *Journal of Management Studies*, Vol. 25, No. 1, pp. 217–237.
Drucker, P.F. (1994), The theory of business, *Harvard Business Review*, September/October, pp. 95–104.
Eggers, F. and Kraus, S. (2011), Growing young SMEs in hard economic times: the impact of entrepreneurial and customer orientations—a qualitative study from Silicon Valley, *Journal of Small Business and Entrepreneurship*, Vol. 24, No. 1, pp. 99–111.
Felsted, A. (2011), Suppliers feel the pinch as retailers seek better terms, *Financial Times*, November 21st, p. 23.
Geer Jr., J.F. (1997), Brand war on Wall Street, *Financial World*, May 20th, pp. 54–63.
Georgelli, Y.P., Joyce, B. and Woods, A. (2000), Entrepreneurial action, innovation, and business performance: the small independent business, *Journal of Small Business and Enterprise Development*, Vol. 7, No. 1, pp. 7–17.
Gronroos, C. (1989), Defining marketing: a market-oriented approach, *European Journal of Marketing*, Vol. 23, No. 1, pp. 52–60.
Gronroos, C. (1997), Keynote paper From marketing mix to relationship marketing – towards a paradigm shift in marketing, *Management Decision*, Vol. 35, No. 4, pp. 322–339.
Gummesson, A. (1994), Making relationship marketing operational, *International Journal of Service Industry Management*, Vol. 5, No. 5, pp. 5–21.
Gummesson, E. (1987), The new marketing – developing long-term interactive relationships, *Long Range Planning*, Vol. 20, No. 4, pp. 10–20.
Haines, D.C. (2007), Manufacturer and retailer power in retailer response to trade discounts, *Academy of Marketing Studies Journal*, Vol. 11, No. 2, pp. 1–18.
Hamilton. S. (2003), The economies and conveniences of modern-day living: frozen foods and mass marketing, 1945–1965, *Business History Review*, Vol. 77, No. 1, pp. 33–42.
Hills, G.E. and LaForge, R.W. (1992), Research at the marketing interface to advance entrepreneurship theory, *Entrepreneurship Theory and Practice*, Vol. 23, No. 1, pp. 33–59.
Hisrich, R.D. and Peters, M.P. (1992), *Entrepreneurship: Starting, Developing, and Managing a New Enterprise*, Irwin, Boston, Mass.
Holcombe, R.G. (1988), Entrepreneurship and economic growth, *Quarterly Journal of Austrian Economics*, Vol. 1, No. 2, pp. 45–62.
Hunt, S.D., Arnett, D.B. and Madhavaram, S. (2006), The explanatory foundations of relationship marketing theory, *The Journal of Business & Industrial Marketing*, Vol. 21, No. 2, pp. 72–83.
Kirzner, I. (1973), *Competition and Entrepreneurship*, University of Chicago Press, Chicago.
Kirzner, I. (1992), *The Management of Market Process: Essays on the Development of Modern Austrian Economics*, Routledge, London.

Kohli, A.J. and Jaworski, B.J. (1990), Market orientation: the construct, research proposition and managerial implications, *Journal of Marketing*, Vol. 54, pp. 1–13.

Johnson, G. and Scholes, K. (1999), *Exploring Corporate Strategy*, Prentice Hall, Harlow.

Langlois, R.N. (2007), The entrepreneurial theory of the firm and the theory of the Entrepreneurial Firm, *Journal of Management Studies*, Vol. 44, No. 7, pp. 1107–1124.

Larsen, P.T. (2007), Victory formally declared in the ABN tussle, *Financial Times*, October 9th, p. 22.

Levitt, T. (1973), Marketing myopia, *Harvard Business Review*, July/August, pp. 24–47.

Mintzberg, H. (1979), Patterns in strategy formation, *International Studies of Management and Organisations*, Vol. 10, No. 3, pp. 67–86.

Morgan, R.M and Hunt, S.D. (1994), The commitment-trust theory of relationship marketing, *Journal of Marketing*, Vol. 58, No. 3, pp. 20–39.

Reichheld, F.E. and Sasser, W.E. Jr., (1990), Zero defections: quality comes to service, *Harvard Business Review*, September/October, pp. 105–111.

Schumpeter, J. (1950), *History of Economic Analysis*, Oxford University Press, New York.

Sheth, J.N., Sisodia, N.J. and Sharma, A. (2000), The antecedents and consequences of customer-centric marketing, *Academy of Marketing Science*, Vol. 28, No. 1, pp. 55–64.

Shrader, C.B., Mulford, C.L. and Blackburn, V.L. (1989), Strategic and operational planning, uncertainty, and performance in small firms, *Journal of Small Business Management*, Vol. 27, No. 4, pp. 45–60.

Slater, J. (2010), Sonoco meets economic challenges by delivering enhanced value for its customers through strong partnerships, *Global Business & Organizational Excellence*, Vol. 29, No. 4, pp. 6–14.

Urry, M. (2003), Young leader holds on to his dream: mortgage lender's chief executive is happy in his job—but he knows it may not last forever, *Financial Times*, February 10th, p. 22.

4
Assessing Futures

Global population

As recently as the late 1990s, the United Nations Food and Agricultural Organisation (the FAO) and the World Bank supported the view that improvements in food technology would mean that by 2010 the world's food supply would be growing faster than demand (Schieb, 1999). Unfortunately what these forecasts ignored is that increasing per capita income following economic growth in China and India caused these nations to bid up the price of food and oil; thereby further reducing the poorer nations' abilities to feed their populations and access affordable sources of energy.

Pimental et al (1999) concluded that a major crisis would arise in the 21st century, due to constraints over the adequacy of global agricultural resources. Analysis of the cropland required to support the population growth revealed that developing nations would face problems over access to affordable food supplies. One reason for this is poor land management, leading to major soil erosion at a rate never previously observed since data collection began in the 19th century. Of even greater concern is the future availability of adequate supplies of water for human and agricultural use. Although the world's water supplies are probably adequate, the locations of such supplies are extremely unbalanced (Coles, 2003). The need for improved supplies of water is greatest in the developing world, which is facing increasing problems over the adequacy of existing water supplies. Investments in storage capacity, water distribution infrastructure and desalination plants can partially mitigate water shortages. Such solutions will mean that even in Western nations consumers will face ongoing increases in their water bills for the foreseeable future.

Energy costs

China's industrialisation policies are far outstripping the demand for more oil. Brown (2006) estimated that by 2030 China will consume 99 billion barrels of oil per day. Although in 2009 new offshore oil and natural gas finds have been announced in the Gulf of Mexico, Brazil, West Africa and Australia and new on-shore finds in North and South America, it will take some 10–20 years to bring these fields into full production. Hence energy prices, although probably on average still remaining lower than during the energy price spike in 2007/08, can be expected to remain high (Vaitheeswaran, 2007). What is less certain is the degree to which high oil prices and technological advances will permit shifts to more sustainable energy sources, such as wind or solar power, becoming commercially viable propositions and thereby lead to a flattening of or reduction in energy prices.

Wild card event

Case Aims: To illustrate how unexpected events can alter expectations over situations in the macro-environment.

In forecasting future events, there is always the need to allow for a 'wild card' event because an unexpected development can cause a reconsideration of what will actually occur. One such example in the energy sector has been the advent of 'fracking' in the oil and gas industry. The technique involves injecting fluid at high pressure into layers of shale rock where oil or gas deposits are trapped. This injection causes the shale to fracture and gas or oil flows out of the well (Crooks, 2011).

Fracking is being used extensively in the USA and test wells are being drilled in other countries such as the UK and Poland. In the USA it is estimated that fracking has resulted in a major expansion of the country's forecasted gas reserves and prices have already fallen by over 50 percent in 2012. New discoveries are causing manufacturers to reconsider their future plans in relation to the location of their production plants. US Steel has started using natural gas in their blast furnaces, which has lowered energy costs and reduced the company's greenhouse gas emissions. Higher cost savings are being enjoyed by the petrochemical industry, with the costs of ethylene being reduced to the point where US producers are able to manufacture ethylene, the plastic industry's base component, at costs lower than producers in Asia and Europe. This has already caused American car manufacturers to consider sourcing composites and plastics within the USA instead of importing them from overseas. At the moment there are no gas liquification plants in the USA but it seems probable such plants will be constructed. This is because the country is now able to successfully compete in the international LPG market.

Population ageing

The assumption within many economies is that middle-class families with children in the age group 18–49 provide the primary source of demand and generate the majority of revenue for fmcgs. The problem facing many of the Western multi-national fmcgs in their domestic markets is that the average age of the individuals who constitute the middle class is rising, because older people are becoming the dominant age group within many nations (Johnson, 2004). In Europe between now and the year 2050, the proportion of individuals aged 65 and over in the population is forecasted to increase by 65 percent. Associated with this socio-demographic shift, older people now own an increasing share of a country's total wealth. In the USA individuals aged 55 and over comprise 35 percent of the adult population yet control 70 percent of the net worth of all household assets. The problem for branded goods companies is that these older people tend to spend and consume less than their younger counterparts. The implications of this situation is that sectors such as alcoholic drinks, furniture, clothing and DIY, and suppliers of such, will face major problems generating sales (Chaston, 2009). In contrast, sales in categories such as products for ageing skin care, non-prescription medical remedies, home care services and household security systems can be expected to exhibit significant growth.

In the public sector population ageing results in a reverse situation in relation to the demand for services. With people living longer governments face rising demand for state pensions and healthcare (Klaase and Van der Vlist, 1990). Long before the 2008 global banking crisis, observers of population ageing in relation to the funding requirements for healthcare and pensions were already predicting that many governments would face a massive fiscal crisis (Jensen et al, 1995). The need to resolve some Western nations' current public sector deficit has dramatically increased the financial problems facing these countries. Relief may be achieved by increasing charges for free or subsided healthcare services, but in many cases the only viable solution will be to restrict or discontinue service provision. This has two implications; namely (i) new opportunities will emerge for private sector providers, making services available for more affluent pensioners, and (ii) the younger generation will be forced to accept more of the burden for caring for elderly relatives. This latter outcome will reduce the net disposable income that younger families have to spend on themselves. Furthermore most governments will be forced to increase personal taxation of those in work and this will further reduce the spending power of the 18–49 year age group.

Healthcare

Commercialisation of scientific discoveries such as sulphonamides and penicillin in the 1930s revolutionised the effectiveness of medical treatments. In their ongoing search for new drugs that would confer a virtual global monopoly for the treatment of a specific illness, the European and American companies invested vast sums into R&D, with annual industry research expenditure estimated to be $30–$40 billion by the late 1980s. The reward was the introduction of a whole range of new drugs such as serotonin inhibitors to treat depression, beta blockers for heart conditions and blood pressure reduction medications. Major drug companies claimed the high prices being charged for these drugs was necessary in order to recover their huge investment in research.

The market reality behind the drug industry's success is that sales have only be supported by employers in developed nation economies funding health insurance for employees and governments covering a large proportion of their nation's healthcare costs. Over 90 percent of the sales of branded pharmaceutical goods are restricted to the world's top 20 economically successful democracies. The growing inability of even these nations to fund their social welfare programmes and private sector employers having to cover ever increasing medical insurance premiums have caused governments to question the ongoing affordability of buying ever more expensive new drugs from the major pharmaceutical firms (Marmor, 1998).

To a certain degree governments can reduce their expenditure on drugs by refusing to authorise the use of new, more expensive treatments and demanding that doctors prescribe generic drugs. The latter are older formulations no longer held under a patent by the originating company. The importance of this latter solution is evidenced by the size of the global generic drug market, which exceeded $30 billion in 2000 (Malhotra and Lofgren, 2004). Savings from use of generic drugs will not close the gap between healthcare service demands and the public sector funds available to deliver these services. Consumers must expect governments to impose means tested cut-offs when determining what services are to be made available to their electorates. Some governments can be expected to pass legislation that forces employers and/or employees to purchase public or private sector medical insurance. Nevertheless such moves are at best only a partial solution. Governments will need to find new solutions to overcome this huge fiscal problem. The outlook for the consumer is the self-funding of a greater proportion of their healthcare costs, higher social security payments or, alternatively, self-funding of a larger proportion of other welfare services which are cut as governments seek to protect healthcare budgets.

Political environments

In order to reduce massive deficits some Western democracies are being forced to reduce public sector spending and to make a large number of public sector workers redundant. The only way these democracies will ever be able to return to historic levels of public sector spending is through a significant upswing in economic growth. The probability of this outcome is extremely low, because the only source of economic growth for the foreseeable future will be the emerging economies such as China and India. As firms from these nations gain an increasing share of total world trade, Western nation firms' sales and profitability will decline, national economic growth will disappear and many people will face a declining standard of living.

The scale and nature of fundamental public sector reform lies in the hands of the politicians. Events in the US Congress in 2011/12 during approval of America's debt ceiling and EU governments' attempts to resolve the Communities' sovereign debt crisis suggest that many of the West's politicians prefer avoiding or postponing major decisions. Presumably this indecisiveness reflects a hope that economic problems are self correcting, allowing politicians to avoid becoming unpopular with their electorates. Indecision could risk voters becoming so frustrated that the minority parties, offering solutions based upon more extreme political values, may begin to attract significant support. In the 1930s this outcome led to the rise of fascist dictators, such as Franco, Hitler and Mussolini, to power in Europe. Currently there are only limited signs that the adverse economic conditions facing some Western democracies will lead to similar extreme political outcomes in the 21st century. Nevertheless support for the extreme right wing politicians in the USA, the popularity enjoyed by the right wing politician Marie Le Pen during the 2012 French presidential election and emergence of extreme right and left wing parties in Greece are all indications that electorates might begin to support more extreme political manifestos as their nations' financial position worsens in the face of deepening austerity.

The early 20th century was a period of social change (Jenkins and Brents, 1989), in some cases as the result of violent revolution. Other Western democracies were able to avoid a complete breakdown in social order (Corner, 2002). Nevertheless lessons learned during the Great Depression caused economists and politicians, by the end of World War II, to identify policies aimed at achieving greater social equality. Initiating Keynsian policies to deliver 'full employment' caused fierce debate. Subsequently academics such as Emmerij (1994) posited that many social security systems, although intuitively appealing, risked causing inflation due to governments' inabilities to sustain funding of the welfare state. He noted that complex systems are needed to manage payments and to avoid the emergence of wide-scale fraud. Emmerij (1994, p. 463) suggested that:

> too much 'social' without sufficient 'economic' leads to bankruptcy and an end to growth; too much 'economic' without sufficient 'social' leads to social unrest and therefore also to an end to growth.

In late 19th century some politicians sought to implement fundamental social reform whereas today many politicians seem to be prepared to modify manifestos to reflect emerging value shifts and the changing nature of society (Birnbaum, 1999). In those democracies where governments are formed through coalitions, one possible explanation for a lack of reforming zeal is the unwillingness of politicians to express views that could fracture their alliances.

Comparative advantage

Since the creation of the first nation states, economic performance has been influenced by the level of cross-border trading. This activity usually reflects one country enjoying some form of 'comparative advantage', permitting the supply of goods at a price or quality which exceeds that of similar output by other nations. Sources of comparative advantage include lower labour costs, abundant raw materials, extensive tracts of land, a climate suited to specific crops, contiguous borders to international markets or a skilled labour force. Comparative advantage is however rarely a static scenario, because factors such as exhaustion of resources, rising labour costs or political change can alter a nation's export capabilities.

Adam Smith and David Ricardo were two economists who promoted the benefits of free trade and the exploitation of specialist capabilities to support greater export activity. Adam Smith's perspective was that where nations excelled in all areas of economic activity, then these nations should exploit every activity to accrue wealth through exporting. David Ricardo took a somewhat different perspective, arguing that even where one nation has a greater abundance of resources than another less well-endowed nation, cross-border trading should still occur because there is benefit in both nations concentrating on activities where they enjoy a relative advantage.

Despite ongoing debates between economists over the most relevant theoretical models to explain cross-border trading, within posited alternative propositions the Ricardian idea of comparative advantage still remains the core idea for explaining international trade. Currently 'new trade theory' is extremely popular. This concept focuses on the positive role that new or changing technology can have on the flow of goods between nations. The proposed theory is that, to sustain economic leadership, developed nations should give priority in the allocation of scarce resources to their most advantaged industries. This will result in an ongoing increase in both the productivity and social wealth of the nation and the maximisation of national capital.

Many developing economies enjoy the benefits of low labour and access to certain non-renewable natural resources. The problem with globalisation is that capital, labour, resources and other production factors can flow more freely between nations. This allows the substitution of capital for labour, work force up-skilling and technology advances that permit new materials to become replacements for natural resources. Hence access to natural resources or low labour costs in today's world may no longer confer long-term, significant comparative advantage. Under these circumstances developing nations will only achieve increased long-term economic wealth by adopting what Wenqiang and Shaojie (2008) described as a 'forging ahead' strategy. This approach is illustrated by economic recovery in Germany and Japan after World War II, which involved creating highly industrialised economies capable of competing in the same market sectors as those in which leading economies have previously achieved high levels of wealth generation.

China and India have clearly learnt lessons from observing the economic gains achieved in world markets by their neighbours', such as Japan and Taiwan, adoption of a forging ahead strategy. India has enjoyed success in the IT industry and in the production of generic drugs. China is successful in areas such as telecommunications, computers and sustainable energy. Both of these nations have the added advantage of large populations. This means that as peoples' incomes increase the growing middle classes will be highly influential on domestic demand, which can further enhance prospects for economic growth.

Market demand is a key driver influencing economic growth and is one of the factors in Porter's (1990) 'Diamond Model' identified as influencing the competitive advantage of nations. This is because a high domestic demand is a catalyst that drives firms towards higher efficiency and productivity in their domestic market. The intensity of domestic competition can assist these firms to become more globally competitive. The second element of Porter's model is 'factor conditions', which he classified into five major categories: human resources, physical resources, knowledge resources, capital resources and infrastructure. The third element of the Diamond Model is 'related and supporting industries', with the most important of these being suppliers and downstream intermediaries.

The fourth element in the Porter model is 'strategy, structure and rivalry', comprising factors that characterise a firm's market interactions. Growth strategies are usually associated with high competitiveness and are often a driving influence in a firm seeking to remain ahead of the competition. Structure refers to the industry composition in terms of the degree of industry concentration. Rivalry involves both the number of players and the intensity of competition within an industry. Intense rivalry usually leads the most successful firm to develop the capabilities which support greater competitiveness upon entry into overseas markets.

Firms in Western Europe and North America have enjoyed the advantages of domestic market demand, well developed factor conditions and supportive infrastructure. Nevertheless these firms can expect to face increasing competition in overseas and domestic markets from organisations based in emerging economies implementing a forging ahead strategy. Merely remaining in the same markets and offering unchanging products or services will inevitably result in declining sales and profitability. Hence, in the context of Porter's Diamond Model, survival will usually require a strategy which focuses upon innovation to develop next generation or new-to-the-world products or services.

The imperative of focusing upon innovation is critical because in many instances 'first movers' enjoy higher financial performance than later market entrants (Lieberman and Montgomery, 1988). In terms of enjoying a comparative advantage—conditions that are supportive of successful innovation—the developed nations still retain superiority conferred by being the world's leading 'knowledge economies'. This is because these countries (i) still remain the home of major firms who continue to be global players in areas of industry where advanced technology is a crucial attribute and (ii) are where the vast majority of leading research universities and research institutes are located. The critical importance of being a knowledge economy was highlighted by Chen and Dahlman (2005) and Raval et al (2009). They concluded that firms in developed economies seeking to retain global superiority must exploit the comparative advantage of having domestic operations located within the world's leading knowledge economies.

Defence industry trends

Case Aims: To illustrate how the defence industry can still provide a future source of technology innovation in developed economies.

In the past the developed nations' manufacturing sector has been greatly assisted by government spending on defence. Expenditure on equipment has generated a large number of jobs. Government funding of R&D to develop new weapons also has a spin-off effect of creating technology that has subsequent applications in production of non-military goods. For the foreseeable future these benefits are likely to be severely reduced as Western governments partially solve their deficit problems by reducing defence spending. Another problem is that overseas contracts for defence equipment are now often accompanied by the customer demanding an offset contract involving a certain proportion of the contract being undertaken inside their country. In the UK in 2011, BAE Systems announced 3,000 redundancies largely due to a slowdown in production of Hawk jet trainers. Concurrently BAE announced increased activities and new jobs associated with building this aircraft inside India.

Despite government cutbacks the defence industry still has opportunities in development work on products leading to technology spin offs that can be exploited in civilian markets. One example is the growing demand for pilotless aircraft or 'drones'. These machines have proved their worth in counter-terrorism campaigns (Anon, 2011). In terms of the civilian spin offs for drone technology, there is always the problem of the latest military advances having to remain secret. Nevertheless knowledge gained in drone technology offers the aerospace industry new opportunities such as in using civilian drones in dangerous environments (e.g. fighting fires), combating crime and supporting domestic anti-terrorism programmes.

Development work to achieve 'hyper-manoeuvrability' drones for the military will assist the development of new composite materials and advances in avionics. These will be extremely important for companies such as Boeing in retaining their market position in the civilian aircraft industry as firms from countries such as China start moving into this manufacturing sector.

Although total spending on defence equipment in the Western economies will decline over the next decade, one area where expenditure will increase is cyber warfare. The sector has become increasingly important since the advent of the Internet because countries can now attack each other through a range of activities, such as injecting viruses into military defence communications systems, shutting down electricity distribution grids or disrupting production activities such as apparently occurred within Iran's nuclear industry in 2010. Hence the defence industry is involved in a massive expansion of technology for attacking and defending computer systems. Although revenue from cyberspace is still small in proportion to total sales, it is forecasted to be the area of fastest growth for the foreseeable future. Importantly increased R&D can be expected to generate civilian spin offs capable of providing the basis for new innovations in the world's IT, telecommunications and broadcast industries.

Political economics

As understanding of economics has evolved over time politicians have been stimulated to control nations' performance by introducing new economic policies. Frequently these actions reflect a desire to retain the support of the electorate and to be re-elected. Since the 1980s an area of great concern among electorates has been rising unemployment. In the face of increasing public sector deficits in many Western nations, Keynesian policies have become a less viable option because governments have been forced to reduce public sector spending, make public sector employees redundant and decrease expenditure on major new capital infrastructure products.

Without the ability to reduce unemployment through higher public sector spending increased political attention has been given to job creation in the private sector. Some governments have focused on the concept of needing to expand their country's manufacturing sector. This perspective does seem to ignore the fact that globally employment in manufacturing is on the decline (Ward, 2006). Approximately 30 million manufacturing jobs were lost globally between 1995 and 2002, with two-thirds of those losses occurring in China. This is because manufacturing output per worker has risen dramatically due to productivity gains generated by increased use of ICT, supply chain restructuring and outsourcing. In addition manufacturers increasingly seeking to survive are focusing their efforts within high-tech goods manufacturing sectors which only employ a small number of highly skilled workers.

Some politicians in the developed nations believe their country has become too reliant upon service sector employment. Nevertheless these politicians do need to recognise that once a nation reaches the mid-point in economic development, service industries will inevitably become a more important source of job growth than manufacturing. Contrary to popular belief, a substantial percentage of these service jobs are in high-skill and high-wage sectors. Furthermore the more dynamic a nation's service sector becomes, the faster more jobs will be created—accompanied by an accelerating increase in GDP. This outcome reflects the fact that the developed nations' service sector includes some high wealth-generating industries such as business and professional services in addition to lower paid jobs in sectors like retailing or catering. Some politicians worry about service sector jobs moving offshore. This trend should not permanently lower the nation's level of employment as long as governments are investing in up-skilling the workers whose jobs have been displaced by off-shoring, because the affected individuals can then move into higher value-added jobs in the domestic economy (Garner, 2004).

Among middle and high-income economies, services generate in the region of 60 percent of all employment; and the higher a country's GDP per capita, the higher the share of jobs provided by the service sectors. Bailey et al (2006) pointed out the following myths in relation to service sector employment:

Myth 1: There is little scope for innovation in local services—whereas in reality there are still major opportunities for productivity improvements in service industries like electricity supply and telecommunications.

Myth 2: Increasing service sector productivity will rapidly increase unemployment—whereas employment in sectors such as retailing and tourism actually increases when the sector implements actions to increase productivity.

Myth 3: Services are a more unreliable source of jobs—it is true that in service segments dominated by small scale operations, frequent job losses occur along with the relatively high failure rates of smaller businesses. However these losses are usually offset by new small firms being created.

Crespi and Pianta (2008) posited that the ability of Western nations to reduce unemployment is heavily dependent upon macroeconomic policies to assist the private sector focus upon the exploitation of innovation in order to sustain a competitive advantage over firms based elsewhere in the world. This perspective was supported by these researchers' study of competitiveness in Europe. This demonstrated that firms which focus upon retaining leadership in technological competitiveness can generate new jobs from product innovation and entry into new markets.

Wealth generation from knowledge-based services

Case Aims: To illustrate how services can provide an effective source of wealth generation in the developed economies.

When ARM Holdings, a Cambridge-based firm, was established it is unlikely anybody forecasted that in 20 years time the company's processor architecture would be the first choice for use in leading-edge products such as smartphones (Edwards, 2010). From the outset the founders recognised that processor architecture based upon reduced instruction set computing (RISC) would eventually enhance the capabilities of mobile electronic devices by overcoming the problem that conventional processors are too large, too complex and too slow. A key event occurred when their design attracted the attention of Apple Computers who eventually entered into a joint venture to develop the first generation of commercially viable technology based around the ARM processor.

The company's management was sufficiently astute to recognise that they would never have the financial resources to move into the manufacture of microchips on a scale which would permit ARM to compete with giants in the industry such as Intel. The identified alternative strategy was to sell technology licences to larger companies in the IT industry. This was despite the fact that initially many in the industry felt the concept of a licensable microprocessor was a non-viable one. The breakthrough occurred when Nokia licensed the ARM architecture for their new GSM handsets and Sharp adopted the technology to power their Nintendo game devices.

Politicians who argue that economic recovery demands a major expansion in their nation's manufacturing base, promoting further growth in the service sector, do need to be aware of past and current industrial trends. Although the Western IT industry was very successful in the manufacture of computers and microchips over time these became sectors where product commoditisation eventually eroded profitability following the entry of Taiwan and China into these industries. Many of the Western industry survivors were firms which evolved in service operation, focusing upon R&D to develop new technologies that can be licensed to the major manufacturing companies elsewhere in the world. Davis (2008) described firms specialising in the generation and licensing of intellectual property (IP) as 'intellectual property vendors'. He noted that these organisations do not engage in the manufacture and marketing of tangible products. Instead their business model is based on licensing the rights to their inventions to other firms, who use this knowledge to produce new electronic products. The risk for the IP vendor is that the customer might attempt to steal the knowledge made available under the licence. A certain level of protection is available through the use of patents. Unfortunately patents only provide a limited degree of protection from knowledge theft. As a consequence firms such as ARM must develop a close working relationship with potential customers, to the point where the latter recognise the major costs–benefits associated with sustaining an ongoing relationship with the IP vendor. Once reciprocal knowledge sharing is accepted as being in the best interests of both parties, close involvement in customer's R&D activities to develop a new product provides the IP vendor with incremental knowledge that can be incorporated into their next generation of technological designs.

Davis proposed that the IP vendor industry can be expected to grow and therefore offer major wealth generation from the service sector in developed nations. He suggested several factors that contribute to this growth. Firstly many large companies have been forced to reduce costs and in part this has been achieved by cutting back on in-house R&D staff activities, causing such firms to rely on external sources to acquire the IP needed to sustain new product development activities. Secondly as global markets have become more competitive major firms have been forced to focus on sustaining the further development of their primary core capabilities, which necessitates going outside the organisation when there is need to access new technology about which there is insufficient internal understanding. The third factor is the recognition that through the use of patents knowledge-based firms can achieve a strategic advantage that protects them from attempts to launch 'me too' propositions by companies based in the emerging economies.

References

Anon. (2011), Flight of the drones; unmanned aerial warfare, *The Economist*, London, October 8th, p. 30.
Bailey, M., Farrell, D. And Remes, D. (2006), The hidden key to growth, *The International Economy*, Vol. 20, No. 1, pp. 48–55.

Birnbaum, N. (1999), Is the Third Way authentic? *New Political Economy*, Vol. 4, No. 3, pp. 437–446.
Brown, A. (2006), The robot economy: brave new world or return to slavery, *The Futurist*, Vol. 40, No. 4, pp. 50–56.
Chaston, I. (2009), *Boomer Marketing*, Routledge, London.
Chen, D. and Dahlman, C. (2005), Knowledge for development program, World Bank Institute, Washington, D.C., October 19th, pp. 20–31.
Coles, C. (2003), The growing water crisis, *The Futurist*, Vol. 37, No. 5, pp. 10–12.
Corner, P. (2002), The road to Fascism: an Italian Sonderweg?, *Contemporary European History*, Vol. 11, No. 2, pp. 272–295.
Crespi, F. and Pianta, M. (2008), Diversity in innovation in Europe, *Journal of Evolutionary Economics*, Vol. 18, pp. 529–545.
Crooks, E. (2011), New wells to draw on, *Financial Times*, October 6th, p. 11.
Davis, L. (2008), Licensing Strategies of the new intellectual property vendors, *California Management Review*, Vol. 50, No. 2, pp. 6–30.
Edwards, C. (2010), 20 years ago today: the making of ARM, *Engineering & Technology*, November 13th, 66–70.
Emmerij, L. (1994), The employment problem and the international economy, *International Labour Review*, Vol. 133, No. 4, pp. 449–67.
Garner, A.C. (2004), Offshoring in the service sector: economic impact and policy issues, *Economic Review—Federal Reserve Bank of Kansas City. Kansas City: Third Quarter*, Vol. 89, No. 3, pp. 5–37.
Jenkins, J. and Brents, B.G. (1989), Social protest, hegemonic competition, and social reform: a political struggle, *American Sociological Review*, Vol. 54, No. 2, pp. 891–902.
Jensen, S., Hougaard, E. and Nieksen, S.B. (1995), Population ageing, public debt and sustainable fiscal spending, *Fiscal Studies*, Vol. 16, No. 2, pp. 1–20.
Johnson, R. (2004), Economic policy implications of world demographic change, *Economic Review Federal—Reserve Bank of Kansas City*, Vol. 89, pp. 39–65.
Klaase, L.H. and Van der Vlist, J.A. (1990), Senior citizens a burden? *De Economist*, Vol. 138, No. 3, pp. 302–321.
Lieberman, M.B. and Montgomery, D.B. (1988), First-mover advantages, *Strategic Management Journal*, Vol. 9, pp. 41–58.
Malhotra, P. and Lofgren, H. (2004), India's pharmaceutical industry: hype or high-tech take-off? *Australian Health Review*, Vol. 28, No. 2, pp. 182–194.
Marmor, T.R. (1998), Forecasting American healthcare: how we got here and where we might be going, *Journal of Health Politics*, Vol. 23, No. 3, pp. 521–542.
Pimental, D., Bailey, O., Kim, P., Mullaney, E., Calabrese, J., Walkman, L., Nelson, F. and Yao, X. (1999), Will limits on the earth's resources control human numbers? Environment, *Development and Sustainability*, Vol. 1, No. 1, pp. 19–39.
Porter, M. (1990), *The Competitive Advantage of Nations*, The Free Press, New York.
Raval, D., Subramanian, S. and Raval, B. (2009), Managers as champions of national competitiveness through strengthening knowledge infrastructure, *Southern Business Review*, Vol. 34, No. 1, pp. 37–50.
Schieb, P. (1999), Feeding tomorrow's world, *The OECD Observer*, Summer, pp. 37–41.
Vaitheeswaran, V.J. (2007), Oil, *Foreign Policy*, Vol. 163, pp. 24–29.
Ward, W.A. (2006), Manufacturing jobs 2005-2010, *Economic Development Journal*, Vol. 5, No. 1, pp. 7–15.
Wenqiang, G. and Shaojie, Z. (2008), Study on forging ahead strategy of developing country from the perspective of comparative advantage, *Canadian Social Science*, Vol. 4, No. 1, pp. 16–18.

5
Tomorrow's Markets

Market change

In terms of determining how the new austerity might alter the demand for an organisation's products or services, a possible starting point is to comprehend what influences customer needs and how these change over time. Ultimately the demand for virtually all goods, with the possible exception of certain areas of public sector expenditure, is determined by consumer demand. In some cases, such as clothes or housing, this demand is of a direct nature, reflecting actual consumer expenditure, whereas in the case of business-to-business (B2B) markets demand is 'derived'. This occurs because sales are ultimately determined by consumer purchasing further downstream within a market system.

Accurate forecasting other than in periods of high economic stability can often be more of an art than a science. Hence the current status of the global economy means it is extremely difficult to be certain about whether demand might change as Western consumers react to living in a world of lower incomes and higher unemployment. This is because (i) there are problems associated with predicting how deep and how long this period of new austerity might last and (ii) reliance on previous periods of austerity, such as earlier recessions, as indicators of the future is complicated by the never ending nature of changing lifestyle shifts and how these influence consumers' spending priorities. Nevertheless, on the basis of retail sales since the onset of the current recession, it is apparent that consumers in most developed economies are changing their shopping habits, having accepted that they are living in a time of greater austerity (Anon, 2011). Fewer people are engaging in impulse buying: most people are increasingly rational about their purchase decisions. Furthermore there are indications that consumers are spending more time considering alternative goods as well as exhibiting a higher level of cynicism over suppliers' promotional claims concerning savings or product quality.

> ### Learning from the past
>
> *Case Aims: To illustrate how consumer spending patterns change over time.*
>
> There are very few markets where demand remains static over time. Factors of influence include product availability, prices and consumption patterns. In relation to consumer consumption patterns, these are strongly influenced by prevailing lifestyles. As consumer incomes change this is accompanied by a shift in lifestyles and product usage. This effect can be illustrated by comparing data from the UK's Annual Family Expenditure Survey for 1956 with that for 2006 (Cohen, 2008). Fifty years ago Britain had only just emerged from post-war austerity, and at the time people spent so little on leisure activities this area of expenditure was not even mentioned in the inaugural survey. In 1956 only 9 percent of total family expenditure covered the cost of housing, whereas by 2006 the figure had risen to 19 percent of total spending. This change reflects the influence of both much higher house prices and the fact that many more people now own instead of rent their homes.
>
> The biggest rise over time however has been the expenditure on energy. The average British family now spends half their total income on energy consumption. Although higher world energy costs are a significant influencing factor, the increase in relation to the proportion of the family budget primarily reflects major lifestyle changes. Fifty years ago few people lived in centrally heated houses and most could only aspire to own a television, refrigerator or washing machine. This has all changed with electrical appliances now being seen as standard items to be found in the majority of households. To this mix has to be added the home computer and mobile electronic devices. In terms of energy costs associated with travel, the proportion of total expenditure has remained almost unchanged, but the mix is very different. In 1956 expenditure on travel at 14 percent of total expenditure was virtually all on public transport, reflecting a much lower level of car ownership. Nowadays over 75 percent of UK households own a car. As a consequence motoring now consumes 14 percent of household spending and spending on public transport now only represents 2 percent of total expenditure.

Shifting behaviours

On the basis of previous economic downturns what can be expected of consumers, especially those in the middle classes, will be to reduce expenditure of leisure activities such as eating out or going on expensive vacations (Kumcu and Kaufman, 2011). This behaviour shift became apparent among American consumers very early into the current economic downturn in relation to the food service industry as evidenced by sales to consumers, governments, businesses and non-profit organisations dropping from $533 billion in 2006 to $513 billion in 2009. Spending levels varied quite considerably across the sector. Full-service restaurant sales declined by 4.5 percent over this period, but only fell by 2.6 percent in fast food outlets. Food sales also declined across other segments between 2006 and 2009, including a decline of 8.8 percent at hotels and motels and of 7.3 percent for purchases from vending machines.

Consumers are exhibiting greater rationality in their purchasing behaviour, with shoppers taking advantage of in-store sales, special promotions and money off coupons. There are also indications of more careful budgeting such as substituting items for comparable but lower cost goods, seeking out stores that offer the lowest prices and a higher proportion of purchases being retailers' own private-label goods. American consumers have also begun to forgo convenience to reduce their household spending. Many consumers are changing their eating habits as indicated by a decline in the purchase of prepared foods and a return to cooking more meals from scratch using basic ingredients.

Research by ISI Chicago concluded that the new austerity has caused US consumers to become more self reliant and take greater care over how they spend their money (Hamstra et al, 2009). The company reported that supersaver and dollar stores have enjoyed an increase in the number of households using their outlets and that consumers doing their weekly grocery shop in supermarkets are influenced by what items are on promotion. Discretionary spending has declined and fewer consumers appear to making impulse purchases.

Drug retailers' healthcare departments have enjoyed a sales increase. This is thought to reflect more consumers engaging in the practice of self care to minimise their medical bills. In contrast beauty and personal care sales growth in these outlets has declined. There is also evidence that shoppers are moving towards using bulk buying to save money on a per unit basis and reducing the number of store visits. Sales for lunchbox fillings have increased as have sales for foods used for at-home entertaining such as crackers and wine.

Some sociologists believe the new austerity has already led to a fundamental shift in consumer attitudes and these will remain in place even when the Western economies start to recover. In the USA some firms have already responded to the emergence of austerity shopping among consumers. Sears, the giant American retailer, has revived a savings plan used many years ago known as the 'layaway programme'. With this scheme the consumer can make a down payment on an item. The price of the item then remains unchanged for a defined period while the individual acquires the rest of the cash needed to complete the purchase (Anon, 2009). Hyundai, the South Korean carmaker, announced that car buyers could return their new vehicle without incurring a penalty if they lose their jobs. This prompted Ford and General Motors to announce a willingness to reduce payments on car loans and leases for a limited period for customers who have lost their jobs. There is also growing evidence that the sub-prime mortgage crisis and the subsequent banking crisis have led to much higher distrust of 'big business' among consumers. As a consequence an increasing number of consumers are placing reliance upon recommendations from friends when reaching a purchase decision, often by relying on social network sites such as Facebook.

In the past people had a tendency to change their behaviour during an economic downturn and then gradually, over time, revert back to their pre-recession purchasing pattern. The degree to which this occurs is influenced by the level of shock caused by economic change. There would appear to be evidence to suggest the level of shock caused by the onset of the current downturn has been sufficient that consumers will not revert to prior behaviour even when indications emerge that the economy is showing signs of recovery. Some market research studies are suggesting that consumers having moved down market by purchasing economy brands perceive these products to offer better value and have little inclination to switch back to their previously purchased premium brands. This new pragmatism is also leading consumers when purchasing durables to focus on basic features and low price in place of wanting to own the latest, most technologically advanced products (Anon, 2010). A permanent shift however to lower priced products is not evident in all sectors. Dibaji et al (2010) reported that in the USA, for certain retail categories such as diapers, trash bags, certain consumer tissue products, batteries, cat litter and many beauty products, 'trade-back-up' is evident. These researchers concluded that these are categories where the suppliers of the lower priced goods are offering products which are of a much lower quality or deliver poorer performance than the more expensive brands in the same category.

In terms of probable changes in behaviour that are becoming evident in a period of austerity, Alioto (2009) suggested that these will include:

(1) Consumers are purchasing less; reflecting the philosophy of 'less is more'.
(2) Product and services will no longer be able to rely on reputation but instead demonstrate definite value.
(3) Consumers are using sources, such as the Internet, to conduct their own research on the actual benefits and prices of product and services.
(4) Personal values are shifting away from 'self' towards needs of others, such as family.
(5) Consumers are becoming more patient about being prepared to delay purchasing high-priced goods and services.

Alioto proposed that these changes need to be recognised by the marketer in order to ensure future marketing strategies are reflective of new attitudes and behaviours among their customers. This will require increased market research activity to identify new trends as soon as these begin to emerge. Furthermore merely relying upon revising promotions or prices is usually an inadequate response. Instead organisations will need to focus on innovation to offer improved and new products or services more suited to meeting the needs of consumers during what is expected to be a long period of austerity.

A somewhat unaffected market

Case Aims: To illustrate that even during an economic downturn the demand for products and services among the more wealthy in a population tends to remain unchanged.

Down through the ages, whatever the state of an economy, there will always remain a small minority who are much wealthier than the rest of the population. The source of this wealth can be from the spoils of war, inheritance, business success or engaging in criminal activities. In some cases the line between business success and criminality can be somewhat blurred, as exemplified by the somewhat questionable activities of certain Russian oligarchs and US banking executives engaged in the sub-prime mortgage market debacle. No matter the source of their money, an important attribute of wealthy individuals is their tendency to sustain a high level of spending whatever the prevailing economic conditions.

This scenario has remained true even during the latest downturn. Hence one way for an organisation to avoid declining sales during a downturn is to focus upon being a supplier to wealthy consumers. In the USA, for example, where consumer spending represents approximately 70 percent of the total economy, it is estimated that 40–50 percent of total consumer spending is by the 20 percent of people who constitute the wealthiest group in society (Chandra and Feld, 2011).

Similar spending trends among the wealthy have been reported in Europe. One indicator of this trend is luxury travel, which generates approximately 25 million trips a year and accounts for nearly 25 percent of total sector spending. The luxury travel market is constituted of high net worth individuals whose average annual income ranges from $1 million to $10 million. Hotels in this sector of the market rely heavily on excellent service and innovative ideas to differentiate them from the competition. Recently a major focus has been on the expansion of personalised services to meet guests' individual requirements. Examples of service provision include valet parking, 24-hour concierge, pillow menus, complimentary limousines, laundry, dry-cleaning and shoe-shining services (Kiessling et al, 2009).

An identified aspect of the very wealthy is a tendency to purchase 'status' goods. This purchasing behaviour is reflective of people striving to improve their social standing through the conspicuous consumption of products and services that confer and symbolise a certain status within their social group. Eastman and Eastman (2011) concluded that status purchasing is only undertaken by a minority of consumers and these tend to wealthier individuals whose personal confidence remains high even when faced with an economic downturn. In contrast less status-conscious consumers, who are the vast majority of a population, tend to perceive the purchase of status products during an economic downturn as an unacceptably frivolous activity.

Heaney et al (2005) concluded that status purchases will be influenced by (i) whether other people will approve of the purchase and (ii) the opportunity the purchase provides for causing others to gauge the material success of the buyer. Eastman and Eastman concluded that consumers with the strongest motivation to consume status goods are less price and value conscious when compared to the average consumer. There is also a significant positive relationship between status consumption and level of brand consciousness, both in terms of brand name and the view that a higher price indicates higher quality. Status purchasers tend to be more brand-name conscious and more likely to see the high prices charged for items such as fashion goods as indicative of superior quality.

Determining strategic response

Although the current downturn is very different from anything previously experienced by Western nations since World War II, Quelch and Jocz (2009) posited that one can still learn lessons from the behaviour of firms in previous recessions as the basis for guiding marketing strategies during this current period of austerity. In their view the biggest mistakes firms can make is to cut costs across all areas of their operations without previously deciding how such actions could jeopardise losing the support of the organisation's most loyal core customers. Hence prior to any hasty decisions over revising spending budgets there is a need for market research to assist in determining the nature of future buyer behaviour.

Quelch and Jocz noted that many consumer businesses were reliant over the last two decades on consumers who believed that they would always be in a position to pay off their accumulated borrowings and were made to feel more confident by an apparent never-ending increase in the value of their homes. Marketers reinforced this belief by promoting the idea that the 'good life' and happiness could be achieved by maximising expenditure on material goods. Once people started losing their jobs or found they were unable to meet their mortgage payments or pay off their credit cards, consumer confidence plummeted and many consumers finally realised that they had been 'living beyond their means' for years. Quelch and Jocz suggested consumer markets could be segmented into the following four basic groups:

(1) *The slam-on-the-brakes* segment who feel most vulnerable and most affected financially by the downturn. Their response is to reduce all forms of spending by eliminating, postponing, decreasing or substituting purchases. Most of the people in this group tend to be lower-income consumers.
(2) *The pained-but-patient* consumers who are optimistic about the future but less confident about the near term in relation to their ability to maintain their standard of living. These consumers are often the largest segment in Western economies and will remain confident until they perceive the risk of unemployment. At that juncture they will switch to being members of the slam-on-the-brakes segment.
(3) *The comfortably well-off* consumers who feel they will not really be affected by any downturn. Hence they tend to spend at pre-recession levels but are probably more selective and less conspicuous in their buying decisions.
(4) *The live-for-today* consumers who make no real changes to their lifestyle. Most are typically urban, younger, rent their home and spend on enjoying experiences more than wishing to own material goods.

In terms of assessing customer behaviour it is necessary to determine the nature of the benefits being delivered by products or services. One approach proposed by Quelch and Jocz is to classify goods into the following types:

(1) *Essentials*, which are perceived as necessary or central to well-being.
(2) *Treats*, which are perceived as indulgences that are considered justifiable in the mind of the purchaser.
(3) *Postponables*, which are needed or desired but whose purchase can be postponed.
(4) *Expendables*, which are perceived as unnecessary or unjustifiable.

Even during periods of austerity loyal customers are the most critical source of reliable cash flow. Hence organisations do need to sustain the level of marketing spend deemed necessary to sustain sales from this group. Possible ways of retaining these customers whilst avoiding unnecessary marketing expenditure include (i) reducing product complexity, (ii) identifying ways of making goods appear more affordable and (iii) focusing upon ensuring the company's products or services are totally trusted by customers. Quelch and Jocz recommended that near-term opportunities will be influenced by the types of customers who are the major purchasers. Prospects are reasonably good for value-based essentials which will appeal to slam-on-the-brakes consumers who are prepared to forgo premium brands in favour of lower prices. These same products will also be of appeal to the pained-but-patient group. Repair services can also be marketed to this latter group because they will be interested in prolonging the life of their existing durable goods. It is very probable however that even when the worst of the current downturn is over customers may not return to their old purchasing patterns. Market research should be undertaken into whether customer behaviour has permanently altered and how innovation might be utilised to exploit this behaviour shift.

Flatters and Willmott (2009) concluded that the new austerity has caused consumers to exhibit an increased level of new thriftiness and a greater desire for simplicity. In their view Western consumers can no longer be expected to purchase goods such as interesting gadgets or electronic products offering the latest technology. Furthermore the concept of being able to afford to be socially conscious, such as being willing to pay a premium price for products made from sustainable raw materials, will decline. Based upon consumer research in both Europe and America, Flatters and Willmott posited that the following issues will become important and thereby influence customer behaviour for the foreseeable future:

(1) Return to simplicity in which product functionality and ease of use will replace the desire to own products offering the latest advances in technology.
(2) Growing demands for more ethical behaviour by large corporations, especially those in the financial services sector. This attitude will be reflected by more consumers being willing to punish organisations which they feel are exhibiting an inadequate level of corporate governance through actions such as switching to an alternative supplier.
(3) Increasing emphasis on being more parsimonious in the selection of products and services and a decline even among affluent consumers to engage in conspicuous consumption.
(4) Promoting more traditional values to their children by exhibiting thrift and greater conservatism in the way they spend money on themselves and other members of their families.
(5) Lower willingness to tolerate poor product performance or inadequate service quality accompanied by a more rapid switch to organisations perceived as being more committed to meeting the needs of their customers.

(6) More realist attitudes towards environmentalism, being prepared to retain a concern about issues such as choosing sustainable products but including in the purchase decision the affordability of being 'green'.

The coupon comeback

Case Aims: To illustrate how austerity has begun to influence consumer use of discount coupons.

Stark evidence of the impact of the current downturn is evidenced by a reversal after 14 years in the redemption rate for store coupons in the USA. This upswing in redemptions reflects both the consumer adopting a more thrifty approach to their weekly shopping and the increased ease of acquiring coupons as a consequence of the social deal sites such as Groupon, LivingSocial and BuyWithMe (Levy, 2011).

Although circulation numbers for the print media have fallen, this has not affected the number of coupons being clipped by consumers. Hence this traditional medium still remains extremely important for brands wishing to distribute coupons to their customers. However to improve redemption rates, and to more precisely target consumers, many companies are now using free standing coupon inserts and others are placing greater emphasis on using direct mail or themed coupon pages.

The volume of coupons being acquired through digital channels such as coupon sites, social media and e-mail is growing at a faster rate than acquisition via the print media but the volume of coupons acquired via digital channels still remains a relatively small percentage of the total coupons redeemed. Part of the reason is that only the minority of the population and companies which offer coupons are using the digital media. The lower use of the digital media reflects the fact that marketers are still seeking to understand how best to exploit the behaviour of the online coupon collector. Some companies also remain concerned about counterfeiting. As time passes however marketers are gaining knowledge of the use of sites such as Facebook and linking consumer profiles to purchase behaviour. This is expected to be reflected in increased use of digital channels by more organisations.

Advances in mobile technology have the potential to dramatically alter coupon redemption behaviour. For example it is now possible via smartphones to target consumers on the basis of their current location. The technology can allow the consumer to exploit a customer's mobile device's location tracking capabilities to get coupons for goods whilst standing near to the product they are seeking when in a store. However the most important change in the use of online technology is the advent of the social deal sites such as Groupon, LivingSocial and BuyWithMe, which permit consumers having elected to pay a fee to use these sites to access even deeper discounts on products and services. Companies such as Groupon negotiate large discounts, usually in the region of 50–90 percent off the purchase price, and then offer these deals to subscribers via a free daily e-mail service. Over time consumers are expected to increase their use of online social couponing sites. This is because individual consumers can propose to other members which products and services the social couponing site should feature. The other dimension of Groupon is that because the coupons can be targeted at a consumer's exact location, even small businesses such as restaurants, hairdressers and dry cleaners can now also use couponing as an effective promotional device.

Examining choices

Theodore Levitt's (1960) proposal that firms should ask themselves 'what market are we in' was designed to avoid exhibiting an excessively narrow or myopic view of their market, which may lead to a failure to identify a new potential source of competition, often as a result of the advent of new technology. In the context of the situation confronting today's organisation, possibly an even more important question is 'what market should we be in as customers respond to the new austerity?' In terms of seeking an answer there is a diversity of options that might potentially result from posing this question.

Some firms will determine that there is no major change in future conditions in their markets. One case where this is a possible scenario is where ongoing strong demand for goods is forecasted but there is a scarcity or finite availability of supplies. A good example of this situation is that facing the world's oil industry, who can within reason expect revenue not to be too adversely impacted because any fall in sales in developed economies will be compensated by increasing demand from the world's major emerging economies. Another possibility of minimal adverse impact is where customers' wealth has not been significantly affected by the current downturn. Hence both their purchase behaviour and price sensitivity remains unchanged. Examples of this scenario exist in markets constituted of individuals who remain wealthy and can be expected to sustain their consumption of goods such as sports cars or jewellery.

The source of revenue for the majority of Western organisations is the purchase activities of the middle classes—the group who will bear the brunt of the impact of the new austerity. Any organisation attempting to sustain demand for the same products in this market sector without making any adjustment in the value of the offered proposition is likely to face a significant decline in total sales. This will occur because competitors have reduced prices or, where this has not occurred, the customer decides the offered goods are no longer affordable and responds by either reducing or terminating consumption. In view of the likelihood of one or both of these outcomes, sustaining revenue will require some form of action to enhance the value of the goods being offered to the customer.

Recent data from the US retail market indicates that consumers are already revising their purchase behaviour in response to their perceptions over actual or future expected earnings The simplest action by a supplier to enhance value in these circumstances is to reduce prices. This trend is already very apparent in many retail outlets in most Western nations. For example, in the USA, one of the first sectors to be severely impacted is the gift market with consumers who earn less than $100,000 expected in 2011 to reduce their spending on gifts by 27 percent versus the prior year (Jopson, 2011). Survey data indicated that one third of consumers will only purchase gifts where the price has been reduced by at least 50 percent, which is why many retailers have had to drastically mark down in-store prices. This strategy has resulted in a major fall in retail profit margins and if sustained over the long term will inevitably lead to many such firms going out of business.

Suppliers may avoid margin erosion through modification of the product mix by introducing a broader range of lower cost items. An example of this strategy in the US retail market is provided by Best Buy, the country's largest electronics retailers. To sustain sales the company has introduced a much broader range of products priced at below $100. Among organisations supplying markets who wish to avoid being forced to accept a lower profit margin on sales to intermediaries the options include (i) identifying internal cost savings with actions such as sourcing lower cost raw materials, or improving productivity, or (ii) modifying their product range to expand their output of lower priced goods (e.g. Amazon's 2011 launch of a lower priced Kindle). Ultimately however such strategies in most cases are easily duplicated by competition and hence should only be perceived as a short to medium-term solution. To sustain long-term survival there is a need to focus on entrepreneurial actions that provide the basis for developing new-to-the world products of appeal to customers whose per capita spending has been reduced, identifying ways of either entering entirely new markets or, alternatively, developing products or services that can provide the basis of creating new-to-the-world markets.

The ostrich management orientation

Case Aims: To illustrate how one group of organisations are refusing to accept that the new austerity will lead to a change in customer behaviour.

Business history is littered with cases of organisations where senior management ignored early signs of market change and insisted on retaining their current strategy to the detriment of their performance over the medium to long term. Examples include Xerox Corporation, which ignored the advent of the low cost, desk top photocopier, Eastman Kodak's continued focus on conventional camera technology whilst others were developing digital cameras and IBM's attempt to retain leadership in computer manufacturing following the market entry by the first generation of PCs.

In view of the frequency with which some organisations ignore the emergence of new market circumstances it can be expected that a number will 'stick their heads in the sand' in response to customer behaviour shifts as a consequence of the new austerity. One early example of this 'ostrich approach to management' has already been provided by many of the UK's universities. In the 1990s, the UK government sought to expand the number of school leavers entering higher education by re-labelling the vocationally orientated polytechnics as universities and subsequently upgrading the status of some further education colleges. In the face of rising costs and the need to reduce the country's public sector deficit, the government announced that universities would be permitted to increase the annual student fee from just over £3,000 to over £9,000 in 2012. This increase, when linked to the living costs of attending university, means that the average school leaver would be faced with loans in excess of £70,000 in order to obtain a degree.

The UK university sector is a two tier system. Within the top tier are prestigious institutions such as Oxford and Cambridge. In the second tier are the ex-polytechnics and the recently upgraded further education colleges. Most institutions have attempted to set their new fees near to or at £9,000. At this level of fees, although top tier colleges can expect to receive a high level of student applications, the same outcome for some of the second tier institutions remains highly questionable. When senior managers from this latter group were asked about their strategy their response was that nothing has changed in relation to school leavers' attitudes towards higher education. In their view the increase in fees and the worsening state of the UK economy will have no impact on recruitment. This is despite the fact that even a year before the fee increases came into effect published market research reports indicated that many school leavers, especially those from socially disadvantaged backgrounds, have decided to seek employment upon leaving school instead of going to university. Further evidence that the decision to behave like ostriches subsequently emerged in data from the country's University Clearing and Application System (UCAS) prior to the start of the 2011/12 academic year. Total applications were down by 30 percent and in the case of degrees where unemployment remains high after graduation, such as media studies or drama, the decline was in excess of 50 percent.

360 degree analysis

Although it is understandable that the primary focus of concern for most managers is the possible shift in behaviour of customers responding to the new austerity, it would be unwise to ignore potential revision in strategy by other players within an organisation's market system. In the private sector, shareholders and the major financial institutions can be expected to apply pressure on company directors to initiate actions to avoid a downturn in organisational performance as a consequence of the new austerity. In turn company directors can be expected to instruct subordinates to 'do whatever is necessary' to avoid a revenue decline. This sequence of events will inevitably lead to some organisations becoming increasingly aggressive in the use of tactics to sustain business performance. As a consequence managers can expect to encounter five potential sources of threat. Hence managers would be well advised to undertake a 360 degree assessment of potential behaviour changes within the market system to determine how best to protect themselves from what can be expected to be increasingly difficult trading conditions.

Of the five possible sources of aggression, the most obvious will be that of other firms operating at the same level within the market system. Some can be expected to implement confrontational actions based upon changes in promotional spending or pricing policy. Similar behaviour can also be expected by suppliers of substitute goods who perceive that the declining standard of living of some of the organisation's customers may force them to switch to a more economical purchase proposition. Thus in the tourism industry, for example, hoteliers should expect more aggressive marketing activities by alternative suppliers of holiday accommodation such as caravan and camping sites.

The most likely serious threat facing Western firms is that posed by new entrants from developing nations which are able to exploit lower production costs to compete very aggressively on the basis of price. During periods of economic stability little known overseas brands usually face severe difficulties overcoming the loyalty of customers to their existing suppliers. New austerity significantly increases customer price sensitivity and this will be reflected in a much stronger interest among customers in switching to less well-known, lower priced imports. China is in a very strong position to exploit the new entrant opportunity (Leahy, 2011). Even in developing nation markets the Chinese are already building market share, as is demonstrated by the success being enjoyed by Chinese car producers JAC and Chery in Brazil. This achievement is occurring despite the fact that Brazil has a large and successful domestic car construction industry.

The nature of the threat from downstream aggressors is likely to be highest when these organisations are in a strong negotiating position. This may occur because there is a significant decline in the number of downstream customers. Thus should the developed nations' airlines decide to freeze their acquisition of new aircraft as passenger numbers decline in the new austerity, for example, this will leave others such as the Middle East and Asian airlines in a much stronger position to demand major price reductions when purchasing new aircraft from Boeing or Airbus Industries. The other source of downstream power can occur because a very small number of intermediaries control the distribution of products within a market sector. An example of this scenario can be seen in those countries where the majority of consumer goods are sold through a small number of national retail chains. Where this occurs these retailers can use the threat of de-listing any suppliers who refuse demands to reduce prices or increase the level of spending to support in-store promotions.

Similar to the possible outcomes associated with the downstream threats, the degree of aggression that can be expected from upstream organisations will be determined by the power and control these can exert within the market system. For example, a supplier of a key component used by the majority of downstream manufacturers will be in a strong position to resist demands from customers to reduce prices, even in those instances where these latter organisations have been forced to cut their own prices to sustain revenue. Where the dominant supplier wishes to sustain sales volume customers may be required to source a greater proportion of their total purchases from this company or risk encountering problems such as delays in goods being delivered or being classified as a low priority customer when the supplier is setting manufacturing schedules. Should a dominant supplier have control over a very scarce or expensive resource, the organisation might also decide to move downstream by acquiring one or more existing customers or creating a new downstream operation. Examples of this scenario are provided by developing country suppliers in the energy industry, such as the Brazilian oil giant Petrobas moving into petrol retailing in other South American countries and the Russian natural gas company Gazprom seeking to enter the UK utility market by purchasing a major power station.

Visioning

For those organisations sufficiently aware of the need to reconsider future actions in the face of the new austerity, the issue of how the planning process can be implemented arises. Unfortunately the classic linear strategic planning model strongly promoted by most business schools for responding to change is rarely applicable (McGrath and MacMillan, 2009). This is because the model is based around posing the question of 'where are we now?' Answering this question merely provides the basis for extrapolating from past experience in order to define future actions. Such an approach will rarely be effective because the planner needs to identify radically new actions which will be relevant in the very different business environments that are associated with the new austerity. Furthermore answering the linear strategic planning model's second question of 'where are we going next?' will be frustrated by a lack of information about what future market conditions are likely to be encountered.

In the face of these constraints, individuals and organisations will need to engage in the process of 'visioning'. The activity is not based upon guess work, although being prepared to engage in intuitive thinking can often provide some very useful additional insights concerning expectations about future events. Finkelstein et al (2008) concluded that visioning works most effectively when grounded in existing knowledge and understanding that a past–present–future approach to thinking and analysis is the optimal approach for evolving the most effective vision of the future. The approach involves critical reflection in relation to each of the four main dimensions of the enterprise; namely organisation, culture, markets and relationships. To be effective Finkelstein et al posited that the vision should be both inward looking (i.e. focused on organisation and culture) and outward looking (i.e. focused on markets and relationships). An effectively managed visioning process should have the effect of stimulating reflexivity, creating discussions, challenging of existing assumptions, questioning long established operational conventions and injecting new thinking into the organisation. Focusing managerial thinking on internal and external business drivers which determine organisational performance can act as catalyst for identifying strategies and actions capable of exploiting the opportunities that will arise as a result of the new austerity.

Millett (2011) expressed the concern that visioning can result in organisations formulating aspirational views of the future based on what managers hope will occur, not actually what is likely to occur. He suggested that a more effective option is for the organisation to engage in 'futuring'. In his view the difference between visioning and futuring is that the latter involves examining what appears to be plausible in the face of emerging trends, evolving conditions and even, possibly, disruptive change. Determination of plausibility is achieved by accepting the premise that the past can provide indications of the future. This is because identifying historical trends may provide the basis for defining patterns and long-term consistencies in market behaviour. Millett accepts that history may not repeat itself, but that nevertheless certain resultant behaviours and outcomes do re-occur. This perspective has led his formulation of the following principles to assist the futuring process:

Principle 1 is that the future will be constituted of combinations of prior events accompanied by what is probably some degree of indefinable change.

Principle 2 is that although futures cannot be predicted with precision, by the incorporation of prior experience, known facts and allowing for a degree of uncertainty, certain probable outcomes can be expected to occur.

Principle 3 is that because futuring and visioning offer different perspectives the combination of these two activities is capable of generating greater insights than merely relying upon only one of the two activities.

Principle 4 is that where forecasts are utilised to examine the future these should be based upon rigorous analysis and a clear definition of any assumptions which have been made in order to generate data of sufficient accuracy to be meaningfully incorporated into any future plan.

Principle 5 is that there is no such thing as a totally accurate forecast and hence there must be a willingness to revise thinking and plans when new information becomes available that indicates forecasting inaccuracies.

In terms of visioning or futuring, McGrath and MacMillan (2009) favour a discovery-driven approach that emphasises searching for the right answers, which can thereby reduce the assumption-to-knowledge ratio. Fundamental strategic change may ultimately prove to be necessary, but where possible most organisations would be advised to in some way to try to sustain the existence of their core business. These authors proposed that this will involve focusing upon (i) initiating the renewal process, (ii) evaluating change options using financial models and (iii) mapping the future growth portfolio.

To avoid ostrich management thinking it is vital that managers do not delude themselves into believing their current business model will remain unaffected by the new austerity. Hence at the start of the renewal process managers must be able to produce adequate evidence that the business will remain unaffected by changing market conditions. One way of achieving this outcome is to test whether the current business model has the capability to sustain a relatively aggressive revenue growth target over the next few years. This activity will usually demonstrate that the current business model and core competences cannot support achievement of such a goal. At this juncture the management team need to assess what other existing or potential development projects can provide the source for sustaining future performance and the nature of the organisational structure required to ensure the identified revised project portfolio can successfully be brought on stream. To guarantee that the projects with the greatest potential for success are selected, McGrath and MacMillan recommend the use of an Opportunity Map which permits comparative assessment of alternative options associated with the two dimensions of (i) level of uncertainty in relation to successful development of required technologies/competences and (ii) level of uncertainty in relation to expectations over market size and growth forecasts.

Innovative medical solutions

Case aims: To illustrate how innovation can offer huge potential to lower the cost and enhance the effectiveness of a medical treatment.

In terms of possibly some of the greatest opportunities for exploiting entrepreneurship to build a significant new business in an age of austerity, some of the greatest opportunities probably exist in the healthcare sector. This is because population ageing will continue to ensure rising demand for healthcare and the financial constraints facing governments and private healthcare insurers will force these organisations to seek new ways of reducing the costs of healthcare delivery.

One area of major concern within the healthcare sector is the rising level of obesity in many nations, which can lead to individuals developing medical problems such as diabetes, heart conditions and circulatory problems. Currently the accepted 'last resort' treatment for the excessively obese is gastric bypass surgery which is a very drastic and potentially risky form of medical intervention. An alternative innovative solution has been developed by the American company Intra-Pace. This is a 'gastric pacemaker' (Leake, 2011). The device, known as Abiliti, works by stimulating the nerves linking the stomach and the brain. This causes an individual to feel full even when they have eaten very little food. For many obese people snacking throughout the day is the cause of their weight gain. The purpose of the device is to prompt people to break this habit and only eat at meal times. The device consists of a sensor embedded in the stomach wall which detects when food enters the stomach. This information is transmitted to a stimulator located close to the vagus nerve which then transmits the fullness message to the brain. In order that people can eat at meal times the device is programmed to switch off for a brief period three times during the day. Currently the cost of the treatment in the UK is £10,000 but as with any electronic technology this cost can be expected to decline over time and make this solution to extreme obesity much more affordable and safer than the alternative of a gastric bypass. As already demonstrated by pacemakers for treating heart conditions, the concept of utilising sensors embedded elsewhere in the human body to lower the costs of other medical conditions clearly represents a potentially huge new market opportunity for existing and new entrepreneurial firms within the healthcare sector.

References

Alioto, M.F. (2009), What post-recession behaviour means for marketers today, *Marketing News*, September 9th, pp. 34–41.

Anon (2009), Business: from buy, buy to bye-bye; consumer psychology, *The Economist*, April 4th, pp. 67–68.
Anon (2010), The post-recession consumer, *Trends Magazine*, February, pp. 16–24.
Anon (2011), Austerity consumer: 'my economy' mindset sparks a new approach, *Marketing Week*, January 13th, p. 32.
Chandra, S. and Feld, A. (2011), Wealthy shoppers buoyed by stock gains are spurring the economic recovery, *BW Magazine*, January, pp. 2–3.
Cohen, N. (2008), Fifty-year journey from austerity to iPods, *Financial Times*, January 29th, p. 3.
Cravens, D.W., Piercy, N.F. and Baldauf, A. (2009), Management framework guiding strategic thinking in rapidly changing markets, *Journal of Marketing Management*, Vol. 25, No. 1/2, pp. 31–49.
Dibaji, A., Powers, S. and Keswari, P. (2010), *The Great Recession and Shifts in Consumer Behaviour*, The Black Book, Berstein Consulting, New York.
Eastman, J.L. and Eastman, K.L. (2011), Perceptions of status consumption and the economy, *Journal of Business & Economics Research*, Vol. 9, No. 7, pp. 9–19.
Finkelstein, S., Harvey, C. and Lawton, T. (2008), Vision by design: a reflexive approach to enterprise regeneration, *The Journal of Business Strategy*, Vol. 29, No. 2, pp. 4–13.
Flatters, P. and Willmott, M. (2009), Understanding the post-recession consumer, *Harvard Business Review*, July/August, pp. 106–112.
Hamstra, M., Veiners, S., Enis, M. and Gallagher, J. (2009), Consumer mindshift, *Supermarket News*, August 27th, pp. 16–24.
Heaney, J-G., Goldsmith, R.E. and Jusoh, W.J.W. (2005), States consumption among Malaysian consumers: exploring relationships with materialism, *Journal of International Consumer Marketing*, Vol. 17, No. 4, pp. 83–98.
Jopson, B. (2011), US retailers tempt cash-strapped shoppers, *Financial Times*, October 29th, p. 17.
Kiessling, G., Balekjian, C. and Oehmichen, A. (2009) What credit crunch? More luxury for new money: European rising stars & established markets, *Journal of Retail & Leisure Property*, Vol. 8, No. 1, 3–23.
Kumcu, A. and Kaufman, P. (2011), Food spending adjustment during recessionary times, *Amber Waves: The Economics of Food, Farming, Natural Resources, & Rural America*, Vol. 9, No. 3, pp. 10–17.
Leahy, J. (2011), China car makers receive a warm welcome, *Financial Times*, October 31st, p. 4.
Leake, J. (2011), Yum, this gastric pacemaker fills me up, *The Sunday Times*, November 6th, p. 14.
Levitt, T. (1973), Marketing myopia, *Harvard Business Review*, July/August, pp. 24–47.
Levy, P. (2011), Cashing in on the coupon comeback, *Marketing Week*, April 30th, pp. 15–18.
Quelch, J.A. and Jocz, K.E. (2009), How to market in a downturn, *Harvard Business Review*, April/May, pp. 54–62.
McGrath, R.G. and MacMillan, I.C. (2009), How to rethink your business during uncertainty, *Sloan Management Review*, Vol. 50, No. 3, pp. 25–33.
Millett, S. (2011), Five principles of futuring as applied history, *The Futurist*, Vol. 45, No. 5, pp. 39–41.

6
Tomorrow's Competences

Contrasting theories

Up until the 1980s, concepts associated with optimising organisational performance were usually based upon the premise that strategy formulation requires developing ways of exploiting the opportunities available within the external environment and that any strategic response is determined by the nature and structure of the industry the organisation belongs to. This emphasis on environmental orientation is exemplified by Porter's (1980) 'contending forces' model. Critics of environmentalism have expressed concern that emphasis on the external market can be detrimental to organisational performance. This is because reliance on a purely market-orientated strategy, without regard to the internal competences necessary to support delivery of products or service performance, may lead to an organisation being overtaken by competitors who have developed more advanced internal competences; thereby being able to offer a superior benefit proposition.

The proposed alternative strategic philosophy, which subsequently became known as the 'resource based view' (RBV), of the firm is based upon the idea that in increasingly competitive markets where all firms understand customer needs differentiation can only be achieved by the organisations focusing upon and exploiting some form of superior internal capabilities (or 'competences'). The core premise of RBV theory is that achievement of a competitive advantage is reliant upon an organisation's ability to organise resources to produce goods and services superior to that of other market participants. The role of management is to define and guide the most effective utilisation of internal critical resources and, in those cases where new capabilities are required, to implement actions to acquire these new competences.

Competence can be defined as an ability to co-ordinate the deployment of available assets to permit an organisation to achieve specified strategic goals. For any aspect of internal operational activity to be recognised as a competence, it should meet the three conditions of organisation, intention and goal attainment. Competence building involves any process which leads to changes in existing assets and capabilities or the emergence of new capabilities that support an improvement in organisational performance. Ownership of a specific competence does not guarantee attainment of a sustainable competitive advantage. This is because two types of internal resources are necessary to establish a competitive advantage; namely assets and competences. Assets are a firm's accumulated resources such as the existence of certain specialist equipment or manufacturing facilities that are necessary to undertake production processes. In contrast competences are the accumulated knowledge and skills which enable staff to undertake the activities that lead to the most advantageous utilisation of the organisation's assets.

The RBV concept acquired academic prominence following Hamal and Pralahad's (1996) proposal that market leaders usually achieve and sustain their business performance by a strategy of consistently exploiting a 'core competence'. These authors posited that an organisation can utilise a core competence to support the development of new and/or improved products and/or enter new market sectors (e.g. Apple's move from computing into electronic communication with the iPod, iPhone and iPad). Alternatively the focus may be directed towards developing superior operational technologies which permit the organisation to compete on the basis of superior performance or price (e.g. the US retailer Wal-Mart which exploited superior capabilities in the areas of procurement and logistics as the basis for out-competing other supermarket chains by offering much lower prices to consumers).

RBV theory would appear to be appropriate in those cases where companies are facing huge difficulties in achieving any form of significant tangible difference in product benefit that can be offered to the market. Under these circumstances identifying a core competence that might provide the basis for a competitive advantage that differentiates the company from competition would appear to be a very logical strategic philosophy. Achieving the aim of delivering a perceivable difference in most service sector industries is usually much more difficult. One potentially successful strategy in this latter sector is to identify a set of competences which permit the organisation to defeat competitors by better value or higher service quality. An example is provided by Singapore Airlines whose competences in delivering high levels of excellent service has resulted in the company consistently being highly rated by international travellers. RBV theory would also appear to be validated by examples from high-technology industries where the key competence is contained within an organisation's ability to assemble a bundle of skills and technologies which permit the development of a unique new technology platform. An example of this outcome is provided by Google's development of the Android operating system, which has permitted manufacturers such as Samsung to more effectively compete with Apple in the smartphone market.

Coyne et al (1997, p. 43) defined a core competence as 'a combination of complementary skills and knowledge bases embedded in a group or team that results in the ability to execute one or more critical processes to a world-class standard'. They proposed that competences can be of two types; namely insight/foresight and frontline competences. These authors proposed that frontline competences tend to be more important in service industries where the quality of an end product or service can vary appreciably depending upon the activities of frontline personnel. In their view insight/foresight enables a company to recognise opportunities in order to develop a first-mover advantage. The source of insights might include:

(1) Technical or scientific knowledge that produces new inventions.
(2) Proprietary data such as information about customers.
(3) Creative flair in inventing successful products.
(4) Superior analysis and data exploitation capability.

In terms of evaluating whether an organisation's core competence is critical for sustaining future performance, Coyne et al suggested that the key questions to be posed include:

(1) Does this competence provide superiority over competition?
(2) Are our skills truly superior?
(3) Is the competence sustainable over the long term?
(4) Does the competence provide the basis for generating superior value for both our customers and the organisation now and in the future?

The original perspective concerning RBV theory was that managerial emphasis should be concerned with the deployment and protection of unique resources and that only limited recognition needs to be given to the issue of whether organisational resources and competences may need to change or evolve over time. Given the increasingly uncertain and volatile nature of global markets, an alternative perspective has emerged: competence-based strategic thinking needs to be more dynamic. This is because an organisation's current distinctive competences may cease to offer the potential for sustaining competitive superiority due to the emergence of fundamental changes in the external markets environment or a sector's production technologies. To avoid this outcome an organisation would be advised to assess the combined benefits offered by current core competence and market positioning, especially given the emergence of a new age of austerity.

Savory (2006) posited that the increasing complexity of markets and technology-based organisational processes demand that new or revised competences must deliver 'higher level capabilities'. To achieve this outcome an organisation must concurrently analyse both core competence and market positioning. During implementation of a dynamic response it will frequently be the case that resources will need to be transferred from one area to another within the organisation. In relation to matching competences to market circumstances Kay (2004) proposed there are five major sources of strength available to the organisation; namely reputation, innovation competence, internal capabilities, organisational assets and external relationships. Kay noted that these strengths will vary from industry to industry and from organisation to organisation operating in the same industrial sector. For example, the technological and engineering strengths necessary to undertake offshore oil exploration and extraction are very different from those required by firms in the long-established iron and steel industry.

Relative competences

In the past Western nations could often survive because their greater competence in the exploitation of technology to make superior products was able to defeat lower price offerings from overseas producers. In recent years however this competence advantage has disappeared in many industries. This is because by learning from developed nation competitors, the technological competence of firms within nations such as China often now equals that found in the developed economies of USA, Europe and Japan. In part this has occurred because Western companies have established their own operations or formed joint ventures with local firms in developing nations leading to the consequent leakage of Intellectual Property (IP) which has accelerated the speed of knowledge acquisition by overseas competitors (Klein et al, 2010).

Even before the onset of the current downturn, Ketchen et al (2007) stressed that, in a world where the only constant is change, entrepreneurship and the effective exploitation of innovation are critical for ensuring the survival of Western corporations. For those managers who reject this perspective of how to sustain existing operations and create new businesses, possibly there is a need to reflect on past tragedies such as that experienced by the American company Kodak. Despite having been the founder of the world's photography industry, Kodak failed to invest in continuous innovation and as a consequence has been overtaken by Asian firms who were quicker to recognise the potential offered by digital technology. As well as a massive decline in the price of the company's shares, company restructuring led to thousands of loyal employees losing their jobs. Another example of a failure to retain entrepreneurial orientation is the Polaroid Corporation, who having created the new market of instant photography, failed to implement a viable digital technology strategy and eventually was forced into bankruptcy.

Leifer et al (2001) posited that the most critical area for sustaining performance is the core competence of innovation. This competence is necessary to permit firms to engage in radical change across areas such as the product, service or internal organisational processes. The aim of such innovation should be to achieve unprecedented superior performance for existing or new products, services or processes which can permit the company to transform an existing market or create new ones. Dahlin and Behrens (2005) suggested that successful radical change should be (i) novel, (ii) unique and (iii) have a significant impact on the technology within an industrial sector.

Rothaermel and Hess (2008) divided innovation strategies into (i) developing and fostering human capital, (ii) investing heavily in R&D, (iii) focusing on developing new products or services and (iv) mergers and acquisitions (M&As). Their conclusion was that the most successful innovation strategy, in terms of impact on performance and increasing the scale of operations, is achieved by those firms that focus on developing and fostering human capital. Rothaermel and Hess believe this innovation strategy enhances an organisation's competence in the acquisition and exploitation of new knowledge. This is critically important because one area where the Western nations *have* retained their leadership is in operating higher education and associated leading-edge research that continues to produce individuals whose intellectual capabilities remain the best in the world. Equally important is the fact that it will take the emerging nations many years before they are able to replicate this educational model.

De Castro et al (2011) considered that the competence of generating new knowledge must be accompanied by specific abilities and behaviours. In their view this combination is required to create a level of 'intellectual capital' that is a fundamental requirement for an entrepreneurial organisation. De Castro et al proposed the components which constitute intellectual capital are:

(1) *Knowledge*, which is the combined influences of (i) formal education, (ii) specific training, (iii) experience and (iv) personal development.
(2) *Abilities*, which are about 'the way of doing things' (or 'know-how') generated by task fulfilment, individual learning and collaboration.
(3) *Behaviours*, which are the actions and traits that cause individuals to act in a certain way; reflecting mental models, beliefs, commitment, self-motivation, job satisfaction and creativity.

Although a common attribute of leading academics and research scientists is a high level of intellectual capital, this competence alone is rarely sufficient to ensure an organisation is able to achieve the level of radical change required to sustain or enhance future performance. This is because there is also the requirement for a capability within the organisation to explore the potential opportunities associated with a radical change, determine an optimal strategy and then implement the necessary actions to convert the identified strategy into a commercially viable proposition (Klein et al, 2010).

In terms of exploiting knowledge to sustain business performance, the major problem facing firms that offer very similar products or services based upon widely available sectoral knowledge is how to differentiate the organisation's offerings from those of the competition. This is not an easy problem to solve at any point in the economic cycle but is one which becomes even more difficult in an economic downturn. This is because customer loyalty is likely to decline. Li and Calantone (1998) posited that the most effective solution to achieving a relative resource advantage over competition is the acquisition of a superior 'market knowledge competence'. These authors proposed that this competence is constituted of the following elements:

(1) *Customer knowledge competence* that enables a firm to explore innovation opportunities created by emerging market demand and reduces potential risks of failing to satisfy changing customer needs.
(2) *Competitor knowledge competence* that provides a diagnostic framework which permits more accurate targeting of the firm's strengths against competitors' weaknesses.
(3) *Marketing–innovation interface competence* that ensures market knowledge is effectively incorporated into internal processes associated with the development of new or improved products.

Price decomposition

Case Aims: To illustrate how a focus on cost minimisation and an entrepreneurial mindset can permit an economy brand to outcompete the competition.

With many of the world's airlines in poor financial shape even before the onset of the new austerity, the European economy airline Ryanair's achievements of a 22 percent year-on-year increase in revenues and 6 percent increase in passenger traffic announced in November 2011 are clearly outstanding (Niththyananthan, 2011). Although from day one Ryanair was positioned as a low cost airline, it was not until the current highly entrepreneurial CEO Michael O'Leary was appointed that the business started to become profitable. O'Leary analysed the operations of the highly successful Southwest Airlines in the USA. As a consequence both companies utilise very similar strategies to achieve the critically fundamental need to minimise operating costs (Anon, 2007). These strategies include:

(1) Sell tickets directly to the customer, mainly via the Internet; thereby avoiding the need to pay commission to travel agents.

(2) Use flexible pricing, charging more for seats on peak demand flights and much less during low demand periods.
(3) Offer a basic travel experience and if customers want added services such as seat reservations or in-flight refreshments they are charged an additional fee.
(4) Direct line of flight to specific destinations, thereby avoiding the more expensive 'hub-and-spoke' model used by many other major airlines.
(5) Only operate one type of aircraft type, which reduces crew and maintenance staff training costs, increases operational flexibility and generates economies of scale for aircraft purchases, costs of parts and lowers servicing costs.
(6) Using smaller lower-cost airports.
(7) Achieving rapid aircraft turn, thereby ensuring planes spend more time in the sky generating revenue.
(8) Providing a single-class cabin, which means the delivery of the same standardised services to all passengers.

Nevertheless there is a fundamental difference between the two airlines. Southwest Airlines continues to operate based upon the performance parameters specified by the company's founder Herbert D. Kelleher. These include exceeding customer expectations in relation to air travel by ensuring on-time arrivals, highly efficient baggage handling, treating customers with respect and where possible staff endeavouring to introduce a little fun into the flight experience. O'Leary appears aware that the less experienced European users of economy airlines have lower service expectations than their counterparts in the USA (Van Pham, 2006). As a consequence Ryanair's strategy is based upon the alternative philosophy of 'price decomposition'. Essentially Ryanair's philosophy is that of delivering the lowest possible seat price. To support this objective the company has carefully examined all aspects of the travel experience and either refuses to provide these or, alternatively, where they are provided will seek to charge an additional fee. For example should a Ryanair flight be delayed, the company does not offer free refreshments and when a flight is cancelled they will not arrange overnight hotel accommodation unless required under EU legislation. The company's usual response to delays or cancellations is to offer a refund or a seat on the next flight, where a seat is available. Another aspect of price decomposition is the company's policy over baggage. Checked-in baggage adds to the number of staff required at check-in and for baggage handling. Additional carrying of baggage adds significantly to fuel costs. Hence to encourage passengers to only use carry-on baggage, the company charges an extra fee for checked-in baggage. Further cost savings are achieved by charging those customers who pay by credit card.

Dynamic competence

The combined influence of changing markets, technology or organisational behaviour will often result in an existing competence being rendered less capable of supporting ongoing financial performance. Teece et al (1997) emphasised that the key role of managers is to lead actions that result in adapting, integrating and reshaping organisational skills, resources and competences. The authors use the term 'dynamic capabilities' to describe this managerial capability. A dynamic orientation requires the capacity to learn and adapt when confronted with new situations or market conditions. O'Driscoll et al (2001) proposed that a failure to reconcile existing competences and acquire new ones may eventually lead a firm into a 'competence trap'. This outcome occurs because the organisation has failed to recognise that changes in market conditions, technology or behaviour of competition have occurred. As a consequence the organisation remains fixated upon exploiting competences which no longer provide the basis for sustaining competitive advantage.

O'Driscoll et al posited that avoidance of the competence trap involves engaging in new knowledge acquisition. This permits recognition of new innovation opportunities and an assessment of whether exploitation of new ideas will require utilisation of existing or totally new competences. Recognition also provides the basis for a renewal of resources, routines, capabilities and core competences. This perspective is reflective of the concept of dynamic capabilities as defined by Teece et al (1997). They had noted that dynamic competence is not necessarily concerned with developing new products or services because actions may be about building, integrating or reconfiguring existing capabilities.

Cavusgil et al (2007) suggested that the car industry is a market environment where ongoing success is critically reliant upon dynamic capabilities. Examples of dynamic capabilities these authors believe have contributed to Toyota's success include:

(1) A total systems approach to product development where people, processes and technology are fully integrated.
(2) Exploitation of the company's understanding of customer-defined values to provide the entry point into any new product development.
(3) Reliance upon continuous innovation that drives all processes associated with the organisation's commitment to improvement.
(4) Technology and product standardisation to support flexibility, reusability, common architecture and processes to minimise 'time-to-market'.

Radical dynamic competence

Case Aims: To illustrate that radical innovation is usually accompanied by the requirement for fundamental changes in organisational competences.

Although radical innovation tends to receive major coverage in the literature, in many cases incremental product or process innovation of an existing dominant technology in an industrial sector can have equal or possibly greater commercial potential. In contrast successful radical innovation will often depend upon the organisation acquiring a completely new knowledge, skills and competences because there are no longer any benefits from continuing to exploit existing competences which have evolved over many years. Although the risks of radical innovation are much higher than for progressive innovation, the rewards can also be much larger, permitting a firm to exploit their competence leadership in order to become the organisation with greater knowledge of a new technology than any of the competition.

An example of the benefits of radical innovation is provided by the UK glass manufacturer Pilkington (Uusitalo and Mikkola, 2010). The glass industry produces two main products, low cost sheet glass and higher quality plate glass. Historically the plate glass production technology has always been well understood with optical performance achieved through extensive grinding and polishing. The high costs of this latter process require an economy of scale and, as a consequence, by the 1950s world demand for plate glass was satisfied by a small number of very large producers.

The competitive dynamics of the plate glass industry were totally changed by Alastair Pilkington who invented a new process known as 'float glass'. In the float glass process a continuous ribbon of glass flows out of the melting furnace, floats on a molten surface and requires no grinding and polishing. The technological problems associated with float glass were formidable. Hence Pilkington decided to ignore developing the technology for the production of low cost sheet glass and focused on developing the process to manufacture high end plate glass for applications such as car windscreens and large plate glass windows for commercial buildings. In the 1960s, by increasing the ribbon speed Pilkington was able to manufacture thinner float glass which permitted entry into the sheet glass industry.

Although Pilkington had a monopoly over the knowledge concerning their new float glass process, the company decided to avoid sectoral chaos by offering to license the new technology to others in the industry. In this way Pilkington gained power in the industry, avoiding a marketing war and the possible intervention of foreign governments seeking to protect their own glass producers.

> Competition in the glass sector intensified in the 1980s following the entrance of US and Japanese glassmakers into the European market and a major expansion by the French company Compagnie de Saint-Gobain. Pilkington was badly affected by downturns in the automotive and building sector markets. The company also purchased Barnes-Hind, which specialised in hard contact lenses, at the point when the market for soft lenses was beginning to grow very rapidly. Pilkington's annual revenue declined from £3 billion in 1989 to £2.6 billion in 1993, and pre-tax profits fell from £300 million to £41 million. The technological competences which propelled Pilkington into a leadership position in the 1960s were not accompanied by the acquisition of the managerial competences required to manage a diversified, multi-location manufacturing operation in a highly competitive global industry. Paolo Scaroni, brought in as CEO in the 1990s to reverse this financial decline, decided the company was employing too many people, operating too many plants and had excessively high overheads (Batchelor, 2000). Pilkington's error, in Scaroni's view, was to become an organisation too reliant upon royalties from the float-glass process and failure to re-invest in developing the competences required to remain an industry leader. A decade of restructuring, investment in innovation, entry into new markets and formation of new manufacturing alliances in emerging markets has enabled the company to become a financially successful business again (www.pilkington.com).

Entrepreneurial competence

Very few organisations have the capability to acquire competence superiority across every area of the activities associated with the production, marketing and distribution of goods (Kyrgidou and Hughes, 2010). Hence management need to determine which areas of competence should be given priority in determining how entrepreneurship will be exploited to sustain future performance (Galunic and Rodan, 1998). One approach to reaching this decision, as illustrated in Figure 6.1, is to determine which area(s) of competence provide the basis for utilising entrepreneurship to remain ahead of competition.

In Figure 6.1 there are two dimensions requiring consideration. One is product/service superiority where the choice is usually between offering superior performance or superior value, the latter most typically achieved in the form of lowest price. The other dimension is operational superiority involving the choice between the production of output or superior competence in supply chain. In many cases, to sustain market leadership the organisation will need to develop an entrepreneurial business model which is based upon

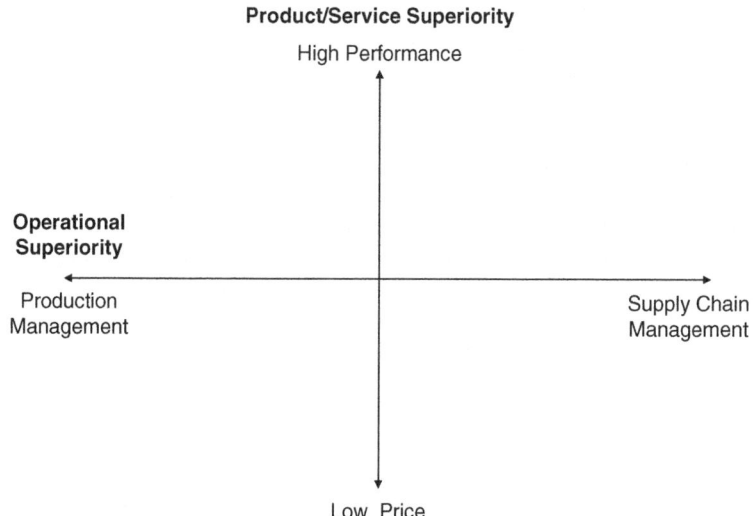

Figure 6.1 Entrepreneurship competence options

a combination of these two dimensions. For example Wal-Mart's success has been based upon a business model of using a highly sophisticated logistics management system to achieve lower operating costs than competition and this superior competence is then exploited by offering lower in-store prices.

Ensuring that dynamic capability results in entrepreneurial behaviour sufficient to retain market leadership through innovation does require an organisational culture which is supportive of transformational change. Menguc and Auh (2006) concluded that positive outcomes from entrepreneurship would be greater in those organisations that are marketing orientated. This is because an understanding of markets can enhance the prospect that development of competences to support innovation is directed towards opportunities which have the highest potential for being adopted by customers. In contrast inadequate understanding of market conditions may result in resources being directed towards a less successful outcome. An example of these two scenarios was provided by the aircraft industry when USA manufacturers perceived the greatest potential in the airline travel market to be the reduction of travel costs; whereas the UK and French manufacturers perceived the market to need reduced travel times, with travel costs being of secondary importance. The outcome was that Boeing developed the 747 (or 'jumbo jet') and the Europeans built the supersonic Concorde. The world's airlines immediately opted for the 747 because this reduced travel costs and expanded the total market for airline travel. Concorde's operating costs were much higher than for a standard jet aircraft and as only a few people were interested in paying the premium fare to arrive slightly earlier, this technologically innovative venture in supersonic travel resulted in the aircraft being a commercial failure.

In addition to correctly identifying desired future competences, Kuratko et al (1990) concluded that senior management's willingness to facilitate, promote, champion and support entrepreneurial behaviour with the allocation of adequate resources to support innovation is also critical. They surmised that senior management's vision must ensure all employees understand and accept that innovation is the organisation's fundamental long-term strategy for optimising future performance. Hornsby et al (2002) expanded this perspective to include senior managers' willingness to provide entrepreneurial staff with autonomy, delegated authority to make decisions, freedom from excessive supervision and removal of restrictive controls over access to needed resources.

Barringer and Bluedorn (1999) posited that an optimal internal environment has to exist in order to facilitate the relationship between the strategic thinking of senior management and the nature of entrepreneurial activities being undertaken within the organisation. These authors concluded there is also a critical need for the organisation to create systems that ensure continuous learning about the external environment, flexibility to revise strategic plans in response to new opportunities and adequate, but not restrictive, strategic controls. Birkinshaw and Gibson (2004) noted differences can arise between existing and new, more entrepreneurial strategies. They proposed that a firm's internal environment must be 'ambidextrous'. This is necessary to enable a firm to switch between explorative and exploitative learning to enable that firm to handle the contradictions that exist between current mainstream activities and future more entrepreneurial actions. To ensure the success of organisational ambidexterity senior managers must present entrepreneurship as the 'dominant logic' within the organisation.

A major problem when extending, modifying and creating dynamic capabilities is the imperfect and intangible nature of the new knowledge, which provides the foundation upon which dynamic capabilities are to be based. Teece (2009) posited that to ensure appropriate knowledge is acquired to develop, re-shape, integrate and re-configure existing and new resources and operational capabilities senior management need to:

(1) Orchestrate the utilisation of complementary assets.
(2) Determine which is the most appropriate business model.
(3) Create the most appropriate internal environment to stimulate and nurture entrepreneurial thinking.
(4) Decide which are the most appropriate strategic investment choices.

(5) Provide the leadership and vision which motivates employees to welcome and to engage in the implementation of change.

Den Hertog et al (2010) proposed that the management of dynamic capabilities to support entrepreneurship is usually more difficult in the case of service firms than for organisations which supply tangible goods. In their view, success in service situations requires effective management of the following processes:

(1) *New need identification* by interacting with clients to determine how new approaches such as service customisation or revised service portfolio can enhance customer satisfaction. These dialogues are more productive when the service innovator is fully informed about how the latest advances in technology permit delivery of different or enhanced forms of service.
(2) *Conceptualising delivery* because delivery of a new service will involve different groups within the organisation or collaboration with others within the service supply chain to link service provision with the service user. In some cases this process may involve the customer who becomes a co-producer or co-designer of the final service.
(3) *Service scaling* reflecting that launching a new service made on a very large scale is often very difficult due to problems over technology capacity, ensuring homogeneous behaviour by employees and the time required to embed new competences into the organisation. Hence there may be a requirement to offer the new service on a smaller scale and postponing the expansion of the service might need to be considered.

Entrepreneurial leverage

RBV theory posits that firms have acquired a valuable, rare, inimitable, non-substitutable capability. Miller (2003) suggested that, even in the case of entrepreneurial organisations, achieving these capabilities is an extremely difficult task. A more realistic objective is to identify strategic 'asymmetries' which are difficult or impossible for competitors to duplicate at an affordable cost. Miller's perspective on possible asymmetric approaches is based upon longitudinal case studies of North American companies, which indicated that identified operational differences are not immediately perceived as the basis for outperforming competition. Over time however, through learning and experimentation, it becomes apparent that one or more asymmetries could provide the basis for more effectively exploiting existing or new market opportunities. Pathways to achieve this outcome include:

Pathway 1 involving customer contacts, experimentation and incremental learning though which the firm begins to perceive that as revisions are made in products, services or operational processes these actions eventually create 'blue water' between the organisation and competition.

Pathway 2 involving reconsideration of effective utilisation of existing resources, introspection and insight causing the firm to recognise that there are more effective ways of exploiting the organisation's competences in satisfying existing or new customer needs.

Pathway 3 involving recognition that weakness, scarcity or poor performance can only be overcome by developing a totally new operational model for serving customer needs. This approach was illustrated by the first economy airlines, such as Easy Jet, which recognised that competing against the existing major players required a very different operational model in order to be able to offer lower prices.

Pathway 4 involving the translating of an apparently unique competence into a viable first mover advantage, followed by ongoing investment in learning and innovation to retain superiority over competition. An example of this pathway is provided by Google, which was an early entrant into the Internet search engine market and, having gained a leadership position, has sustained investment in innovation to develop a diversified portfolio of service propositions.

Another entrepreneurial pathway through which competence superiority may emerge is that of redirecting existing resources and capabilities into what appear to be an entirely new source of opportunity. Miller (2003, p. 971) described this action of extracting additional value from often underutilised resources as 'leveraging capabilities' and proposed that when 'learning takes place, a firm is able to apply the capabilities learned and resources earned in one situation'. He proposed that leveraging can occur in a number of different ways, including:

(1) Exploiting the same competences across different products and industries. For example, the move by a number of Japanese firms in the brewing industry into the production of medical products, involving the large scale culture of bacteria made possible because brewing requires a deep knowledge of bacteria, chemicals, sterilisation, cleaning and testing techniques.
(2) Exploiting existing understanding of customer behaviour to market newly developed products or services to these same customers.
(3) Exploiting segment-related expertise developed with one customer to develop products or services for new customers in the same or related market segment.

Danneels (2007, p. 513) proposed that one of the commonest forms of leveraging 'involves drawing on an existing competence, while using it as a stepping stone to build a new competence'. In his view, where an internal competence is to be leveraged this will rarely prove successful unless the organisation also owns market-related resources such as knowledge of customer needs, preferences and purchasing procedures, distribution and access to communication channels for the exchange of information between the firm and customers.

Danneels (2002) posited that in many cases successful leveraging involves 'de-linking' technological competence from the current product application and then developing the new market management competences to 're-link' the product or service with potential new customers. He concluded that not recognising the need to de-link the product before approaching new customers can result in failure because the firm (i) does not appreciate that the new customers may have very different product needs or, alternatively, (ii) that product technology may require reformulation before the product can actually offer a genuine usage benefit to the new customers. Hence he proposed that during the de-linking process the company needs to (i) recognise technological competence in its own right instead of focusing upon how products are used by current customers, (ii) imagine alternative applications and (iii) define the technological competence required in these alternative applications. Having successfully completed these tasks, re-linking should focus on (i) how to serve new customers utilising the identified technological competence, (ii) building the marketing resources needed to serve the new potential customers and (iii) utilising acquired market understanding to develop new products or services.

> **Sustaining the dynamic**
>
> *Case Aims: To illustrate the importance of sustaining entrepreneurial activities in the face of the new austerity.*
>
> The fundamental changes that can be expected as a consequence of the new austerity may mean that only those firms which continually strive to exploit innovation to respond to emerging opportunities can be expected to sustain their performance. It is probable that the most successful innovations are those which concurrently reduce operating costs whilst sustaining customer value. One example of this orientation is provided by Amazon's move into the creation of local collection centres for products ordered by online shoppers (Harris, 2011). One of the greatest areas of cost for the online retailer, especially in a world of rising energy prices, is the delivery of products to individual addresses. This

> cost is then further increased when the customer is not at home and a re-delivery has to be scheduled. For the customer one of the inconveniences of online shopping is ensuring somebody is at home to take delivery of their purchases. Amazon's innovative solution was the creation of lockers located in locations such as city railway stations, near to where many people work. Upon submitting an order the customer is given the option of a home delivery or delivery to an Amazon locker. Once the delivery has been made, the shopper receives an e-mail containing a unique entry code for the locker. The customer has 24-hour-access to the locker and can collect the delivered goods at whatever time is most convenient to them (e.g. on their way home from work). As the system develops it may even be possible to add further convenience, such as Amazon delivering products to lockers on the day of purchase when goods come from stock in warehouses located near to major cities.

References

Anon (2007), Employees come first at high-flying Southwest Airlines: model contrasts with the Ryanair approach to low-cost aviation, *Human Resource Management International Digest*, Vol. 15, No. 4, pp. 5–11.

Barringer, B.R. and Bluedorn, A.C. (1999), The relationship between corporate entrepreneurship and strategic management, *Strategic Management Journal*, Vol. 20, No. 5, pp. 421–444.

Batchelor, C. (2000), Pilkington hopes to look through glass more brightly, *Financial Times*, March 14th, p. 2.

Birkinshaw, J. and Gibson, C.B. (2004), Building ambidexterity into an organization, *Sloan Management Review*, Vol. 45, No. 4, pp. 47–55.

Cavusgil, E., Seggie, S.H. and Talay, M.S (2007), Dynamic capabilities view: foundations and research agenda, *Journal of Marketing Theory and Practice*, Vol. 15, No. 2, pp. 159–166.

Coyne, K.R., Hall, S.D. and Clifford, P.G. (1997), Is your competence a mirage? *The McKinsey Quarterly*, Vol. 1, No. 1, pp. 40–47.

Dahlin, K.B. and Behrens, D.M. (2005), When is an invention really radical? Defining and measuring technological radicalness, *Research Policy*, Vol. 34, pp. 717–737.

Danneels E. (2002), The dynamics of product innovation and firm competences, *Strategic Management Journal*, Vol. 23, No. 2, pp. 1095–1121.

Danneels, E. (2007), The process of technological competence leveraging, *Strategic Management Journal*, Vol. 28, pp. 511–533.

De Castro, M.G., Delgado-Verde, M., Lopez-Saez, P. and Navas-Lopez, J. (2011), Towards an intellectual capital-based view of the firm: origins and nature, *Journal of Business Ethics*, Vol. 98, pp. 649–662.

Den Hertog, P., Van der Aa, W. and De Jong, W.A. (2010), Capabilities for managing service innovation: towards a conceptual framework, *Journal of Service Management*, Vol. 21, No. 4, pp. 490–514.

Galunic, C. and Rodan, S. (1998), Research notes and communications: resource recombination in the firm, *Strategic Management Journal*, Vol. 19, No. 2, pp. 1193–1199.

Hamal, G. and Pralahad, C. (1996), *Competing for the Future*, Harvard Business School Press, Boston, MA.

Harris, M. (2011), Your new address locker 7, Regent Street, *The Sunday Times*, October 30th, p. 19.

Hornsby, J.S., Kuratko, D.F. and Zahra, S.A. (2002), Middle managers' perception of the internal environment for corporate entrepreneurship: assessing a measurement scale, *Journal of Business Venturing*, Vol. 17, No. 3, pp. 253–273.

Kay, A. (2004), Strategic management, core competence and flexibility: learning issues for select pharmaceutical organizations, *Global Journal of Flexible Systems Management*, Vol. 5, No. 4, pp. 1–15.

Ketchen, D.J., Ireland, R.D. and Snow, C.C. (2007), Strategic entrepreneurship, collaborative innovation, and wealth creation, *Strategic Entrepreneurship Journal*, Vol. 1, pp. 371–385.

Klein, R., de Haan, U. and Goldberg, A.I. (2010), Corporate exploration competence and the entrepreneurial enterprise, *Journal of Knowledge Economics*, Vol. 1, pp. 86–116.

Kuratko, D.F., Montagno, R.V. and Hornsby, J.S. (1990), Developing an intrapreneurial assessment instrument for an effective corporate entrepreneurial environment, *Strategic Management Journal*, Vol. 11, pp. 49–58.

Kyrgidou, L.P. and Hughes, M. (2010) Strategic entrepreneurship: origins, core elements and research directions, *European Business Review*, Vol. 22, No. 1, pp. 43–63.

Leifer R., Colarelli, O., O'Connor. G. and Rice, M. (2001), Implementing radical innovation in mature firms: the role of hubs, *Academy Management Excellence*, Vol. 15, No. 3, pp. 102–113.

Li, T. and Calantone, R.J. (1998), The impact of market knowledge competence on new product advantage: conceptualization and empirical examination, *Journal of Marketing*, Vol. 62, No. 4, pp. 13–29.

Miller, D. (2003), An asymmetry-based view of advantage: towards an attainable sustainability, *Strategic Management Journal*, Vol. 24, No. 10, pp. 961–976.

Menguc, B. and Auh, S. (2006), Creating a firm-level dynamic capability through capitalizing on market orientation, *Academy of Marketing Science*, Vol. 34, No. 1, pp. 63–75.

Niththyananthan, K. (2011), Ryanair lifts profit forecast, *Wall Street Journal*, November 8th, p. 3.

O'Driscoll, A., Carson, D. and Gilmore, A. (2001), The competence trap: exploring issues in winning and sustaining core competence, *Irish Journal of Management*, Vol. 22, No. 1, pp. 73–90.

Porter, M. (1980), *Industry and Competitive Advantage*, Free Press, New York.

Rothaermel, R.K. and Hess, A.M. (2008), Innovation strategies combined, *Sloan Management Review*, Vol. 51, No. 3, pp. 4–9.

Savory, C, (2006), Translating knowledge to build technological competence, *Management Decision*, Vol. 44, No. 8, pp. 1052–1075.

Teece, D.J. (2009), *Dynamic Capabilities and Strategic Management: Organizing for Innovation and Growth*, Oxford University Press, Oxford.

Teece, D.J., Pisano, G. and Shuen, A. (1997), Dynamic capabilities and strategic management, *Strategic Management Journal*, Vol. 18, No. 7, pp. 509–533.

Uusitalo, O. and Mikkola, T. (2010), Revisiting the case of float glass: understanding the industrial revolution through the design envelope, *European Journal of Innovation Management*, Vol. 13, No. 1, pp. 24–45.

Van Pham, K. (2006), U.S. and European frequent flyers service expectations: a cross-cultural study, *The Business Review*, Cambridge, Vol. 6, No. 2, pp. 32–39.

7
Leadership, Vision and Strategy

Visionary leaders

Shelton (1999) and Tischler (2004) both proposed that great leaders have a defined vision of the future, know what they want to accomplish, can define desired outcomes and commit all their energies to accomplishing their vision, often in the face of what may appear to others as insurmountable obstacles. They also concluded that self belief in vision permits these individuals to inspire others to support the necessary actions required to deliver the identified vision.

Morgan (1986) posited that for leaders to be able to gain acceptance for a vision that will energise and motivate others, the vision must be new, credible and comprehensible. Furthermore for a vision to be fulfilled the leader must live their vision and understand that critical importance of gaining co-operation and assistance of other people. Morgan describes this latter process as 'symbolic management'. He noted successful leaders also understand that in some, but not necessarily all, situations there are benefits in seeking the views and inputs of others in order to avoid the terminally damaging outcome of selecting the wrong vision. Morris et al (2005) posited that a willingness to both listen to and sometimes ignore the views of others requires an individual who possesses an appropriate blend of humility and strong personal will (Morris et als 2005).

During the 20th century one such individual was the American president Franklin D. Roosevelt, whose New Deal guided the country out of the Great Depression and continued to provide the nation with effective leadership during the difficult years of World War II. Britain's Winston Churchill was another leader who also fulfilled the role of being an effective leader during this same war. He had a very clear vision of what was required to defeat Germany and their allies, but also recognised that, in addition to enthusiasm and self belief in the correctness of his policies, he needed to persuade others that his ideas were correct. Churchill also exemplified another critically important attribute of a great leader; namely being prepared to generate new ideas and be open to others who can provide new solutions to difficult problems (Longstaffe, 2005).

The phrase 'a man for all seasons' was the title of the play written by Robert Bolt concerning the highly principled Sir Thomas More, the 16th-century Chancellor of England. In real life however the man for all seasons attribute may not always be applicable in relation to many political leaders. This is because their background, experience and skills often mean they are most suited to a specific role at a specific time in history. For example Winston Churchill was an outstanding war-time leader but judged as somewhat less effective in the same role during peace time.

Where visionary political leadership is lacking, this can be detrimental to the economic well-being of an entire nation. Such would appear to be the case in both Europe and the USA following the onset of the banking and sovereign debt crises. In the USA, President Obama, although elected on a campaign platform of promising change and economic recovery, has been unable to persuade the Democrats and Republicans in Congress to reach agreement over actions to re-energise the country's economy. In Europe, Angela Merkel's apparent refusal to accept that political ideology possibly should not obstruct appropriate economic actions has led to a worsening of the sovereign debt crisis, with countries such as Italy and Spain being forced to pay unsustainable levels of interest to refinance their public debt.

As Rowe and Nejad (2009) noted, an unwillingness to act decisively in the face of change is an attribute not unique to politicians. There are numerous cases in the private sector where an organisation's leadership has continued to implement a strategy already rendered obsolete by changing circumstances. In their view, this failure can be attributed to the lack of a long-term vision, inadequate emphasis on innovation and a failure to focus on the acquisition of the core competences required to sustain the organisation in the face of change.

Rowe and Nejad perceived visionary leaders as individuals whose primary focus is on how the future is likely to impact the organisation and how to ensure the organisation is equipped to respond to changing circumstances. These authors recognised that a focus on the future can be risky, because the probability of misreading the future is very high. One obstacle confronting the visionary leader when implementing a new vision is that associated activities may be accompanied by a short-term downturn in financial performance. This outcome can result in the leader facing criticism from external stakeholders, such as the financial community, who have yet to accept there is a need to introduce a new vision in order to more effectively develop and cope with very different future market conditions.

Hickman (1992) suggested that the difference between visionary and conventional management is that the latter tend to focus on the present, and even when these individuals do think about the future that thinking is usually based upon extrapolating past or current trends. A visionary leader's mind resides in the future, with any examination of the present mainly concerned with how a failure to change may have adverse long-term impact. The present merely defines the magnitude of performance improvement that is demanded. Hickman (1992, p. 12) viewed short and long-term performance as being important to the organisation. He commented that 'while the company should not pursue short-term results at the expense of long-term results, neither should it use the pursuit of long-term results to justify poor performance in the short-term'.

Long-term vision

Collins and Porras (1996) proposed that a visionary company is characterised by:

(1) *Clock building, not time-telling*, whereby the company is built upon a vision that can sustain the organisation well beyond the life of any single leader and survive through numerous different business cycles.
(2) *More than profits* is a powerful internal core ideology based upon core values, competences and a sense of purpose, which extends far beyond simply making money.
(3) *Preserve the core but stimulate progress*, which involves protecting core ideology whilst concurrently exhibiting a relentless drive for progress based upon the perspective that change can only occur through discovery, creativity and innovation.
(4) *Audacious goals*, which involve the companies deliberately defining high-risk aims and outcomes, which in some cases may involve 'betting the entire company' in order to retain market leadership.
(5) *Strong culture*, whereby the core ideology is translated into clear cultural and attitudinal patterns that guide the behaviour of the entire work force.
(6) *Shot gun approach to idea generation*, whereby employees are permitted to engage in experimentation, often unplanned or undirected, because this is seen as more likely to produce new or unexpected paths of progress.
(7) *Grow your own*, involving the philosophy that in most cases the best way of creating the next generation of leaders is to select, develop and promote managerial talent from within.
(8) *The mantra of 'good is never enough'*, whereby staff accept the ethic of continuous self-improvement.

Collins and Porras (1995) concluded that at start up only a small minority of organisations can be considered as entities led by a visionary engaged in launching an innovative and immediately successful product or service concept. More typically the organisation will be in business for some years before acquiring the capability to outperform other firms. The researchers' analysis of visionary companies that have enjoyed long-term success indicated these organisations were less likely to have been highly entrepreneurial in their early years and that a significant time period usually passed before a vision emerged that provided the basis for their subsequent entrepreneurial success. Collins and Porras concluded from their analysis of the activities of visionary leaders such as Sam Walton, William Procter, James Gamble, William E. Boeing, Bill Hewlett, Dave Packard and Walt Disney that the success of these individuals can be attributed to high levels of persistence, overcoming significant obstacles, attracting dedicated people to the organisation, persuading staff of the achievability of their vision and being able to guide their organisations through crucial episodes in their respective corporate histories.

Sustaining the vision

Case Aims: To illustrate how one major corporation has sought to sustain a vision which has remained relevant even during periods of significant change.

An example of an organisation which for almost 100 years has remained successful in visioning and creating next generation entrepreneurial ideas is the American firm Boeing Corporation. The vision of the company's leadership team has remained that of retaining market leadership though always striving to improve flight technology. Internal assessment of success is measured in the performance of the company's aircraft and the organisation's reputation for leading-edge technology with financial performance, although important, remains a secondary issue (D'Intino et al, 2008).

Under the guidance of the company's founder William Boeing, the first aircraft were wooden flying boats for use in World War I. After the war life was made difficult as governments reduced the scale of military spending and the market for civilian passenger aircraft remained extremely small. The company self-funded the development of various aircraft for the military and the construction of a civilian aircraft to win a government contract for delivering the mail. The company's innovative B-9 bomber failed to attract military interest but provided the expertise to create the Boeing 247, which was sold to the newly created United Airlines. When TWA and American Airlines then asked Boeing whether they could also develop a smaller, less advanced commercial aircraft, unfortunately Boeing declined this request. As a result TWA contracted with the Douglas Corporation, which went on to develop the DC-1, DV-2 and the highly successful DV-3 (or Dakota).

Egtvedt, Boeing's president at this time, recognised that listening to the opinions and future needs of potential customers could help avoid the company making the type of errors that in essence had handed the civilian aircraft market to the Douglas Corporation. Hence when the Army Air Corp asked for bids for a new multi-engine bomber, Boeing discussed design alternatives with the military before deciding to build a four-engine aircraft. The outcome of Boeing's new design was the highly successful B-17. From the knowledge they gained Boeing then developed the more advanced B-29 (Serling, 1992).

After World War II Boeing realised that jet engines represented the future and declined the Air Force's request for a new bomber design, powered by turboprops. The company was finally given approval to design a jet engine bomber, the B-47. The experience gained permitted Boeing to enter the civilian market with their highly successful Boeing 707. This success was followed by a series of new generation designs such as the 727, 737 and finally, the riskiest of all their development programmes, the Boeing 747 jumbo jet. The latter was a brave decision because industry wisdom at the time was that the next generation of aircraft to be purchased by the world's airlines would be supersonic jets.

By the end of the 1980s Boeing was enjoying a virtual monopoly in the civilian passenger jet market. This position came to an end when European aerospace firms created a competitor, Airbus Industries. Most recently a new battle has ensued leading Boeing to have to make decisions over whether they or Airbus have correctly identified the future of air travel. Airbus has opted for the super jumbo the Airbus 380, which has faced significant delays in being brought to market. Boeing's strategy has been to avoid building an even bigger aircraft and instead, as an interim move, they opted to build their smaller, next generation 747. This is being positioned as offering the world's airlines a solution to rising fuel costs by being as much as 10 percent lighter than previous aircraft, consuming less fuel and delivering a significant seat–mile cost reduction when compared to the A380. In terms of investing in next generation products, Boeing has opted for efficiency over capacity by launching the Boeing 787 Dreamliner which offers outstanding fuel efficiency versus the Airbus 380.

Airbus did steal the march on Boeing in 2010 by announcing the Airbus A320neo, a single-aisle jet aircraft delivering higher operating efficiency accompanied by their plans to offer airlines the opportunity to retro-fit more fuel efficient engines to existing Airbus 320 fleets. Boeing's initial reaction was to remain focused on the development of a next generation version of the Boeing 737. However the company was forced to reconsider their position and, in 2011, Boeing announced the initiation of a re-engineering upgrade of the existing 737 design.

> From their analysis of the reasons for Boeing's long-term success, D'Intino et al (2008, p. 54) concluded that the company provided excellent support for the perspective that in relation to effectively exploiting an entrepreneurial approach to the management of organisations 'success and failure result primarily from internal aspects of the firm largely driven by visionary leadership that not only welcomes change but is also willing to sacrifice financially to incorporate new technologies and develop innovative and highest quality products'.

The purpose of vision

Westley and Mintzberg (1989) suggested that individuals capable of formulating a successful vision usually exhibit the attributes of insight, foresight, idealism, originality and the ability to inspire others. They noted that successful visions are usually the result of a leader's experiences and interaction with specific external environments and events. As a consequence the identified vision is usually the product of a specific point in time and is related to specific events. The purpose of vision is to provide a long-term perspective based upon what is perceived by the leadership to confront the organisation in the future (Kantabutra, 2008). Given the fundamental changes in markets and customer behaviour that are occurring in the current period of prolonged austerity however, the issue arises over whether the new external environment is so fundamentally different from the past that some organisations will be forced to totally re-evaluate the ongoing relevance of their existing vision.

A problem that can arise in advising organisations on the definition and utilisation of a vision is the confusion apparent in the literature over defining what exactly should be covered in (i) the organisation's vision statement and (ii) the organisation's strategy statement. The problem seems to occur because some writers in their definition of mission and strategy use phrases which create duplication of purpose in relation to the two statements (Calfee, 1993). One way of avoiding this problem is to accept Strong's (1997) perspective that the organisational vision should be a vibrant and compelling definition of the organisation's future purpose. In contrast the strategy statement should define the scope of the organisation's activities, covering issues such as products or services offered, markets served and core technologies utilised. Kantabutra and Avery (2010) proposed that some of the attributes for an effective vision include:

(1) Conciseness, clarity, future orientation, stability, challenge, abstractness and an ability to inspire.

(2) Be encompassing of all organisational interests and not a short-term aspiration that can be met and then rendered obsolete.
(3) Relevant to all circumstances and hence unlikely to be altered by market or technological change.

Re-adopting the vision

Case Aims: To illustrate the benefits of recognising the ongoing validity of an organisation's original vision.

Corning Company began to manufacture glass casings for Thomas Edison's light bulbs in New York in the 19th century. However it was the entrepreneurial enthusiasm of Eugene Sullivan in the early 20th century that established the company's vision of exploiting an understanding of the chemistry and capabilities of glass to provide the basis for new innovative products. Included in the subsequent product line were ovenproof ceramics (notably Pyrex and CorningWare), cathode-ray colour TV tubes and fibre optics for voice and data communication (Kelly, 2010).

The huge financial lure of becoming the leading manufacturer of fibre optics for the telecommunications industry caused the company's decision to drop the vision of utilising research on the chemistry and capabilities of glass in order to develop new products. The company became deeply involved in production of fibre optics. Initially this action led to a doubling in annual revenue between 1997 and 2000, and caused the company to expand manufacturing capacity, funded by selling off their large medical services business and the Pyrex and CorningWare operations. In the early 21st century, changes in the telecommunications industry and the entry of other firms into the fibre optics sector left Corning facing excess manufacturing capacity and declining demand.

The board persuaded the retired chairman James R. Houghton, the great-great grandson of Corning's founder, to return as the new CEO. He found the optical fibres group had been receiving the bulk of R&D funds and other areas of research expertise in the chemistry of glass had been left to wither away. Houghton recognised the need to return to the company's founding vision. He drastically reduced the scale of Corning's fibre optics operations and shut down the firm's new small photonics business. He was also, much to his own personal regret, forced to make thousands of employees redundant.

To rebuild Corning, Houghton chose to return to being a diversified, research orientated company and avoid making short-term decisions based purely on financial attraction. Within only a few years Corning has now become a global leader in four market segments:

(1) Display technologies—comprised of glass substrates for LCD flat panel televisions, computer monitors, laptops and other consumer electronics.
(2) Environmental technologies—consisting of ceramic substrates and filters for mobile emission control systems.
(3) Telecommunications—including being the world's largest producer of LCD flat glass.
(4) Life Sciences—comprised of optical biosensors and lab products, for drug discovery.

Corning has remained in the fibre optics industry but now focuses on innovation and development of new products. In 2007 the company launched a new kind of optical fibre, ClearCurve, which is capable of being bent—previously considered to be impossible. The scale of the company's recovery is demonstrated by the fact that in 2006 sales exceeded $5 billion and profits were the highest ever achieved in the company's history. The proven importance of R&D is evidenced by the opening in 2010 of the company's new $300-million research facility at Sullivan Park Research & Development campus in New York State.

Assessing futures

Whether or not the organisation has a formal statement covering the issues of products or services offered, customer target and core technologies, these are issues which do require assessment in terms of how their relevance might be influenced by the new austerity. Even before the current downturn, Cravens et al (2000) proposed that markets are rapidly changing with factors such as innovation, customer diversity, aggressive global competition and extensive access to information all accelerating this change. They posited that market complacency, in which the organisation assumes the future will mirror the past, is not only myopic but also risks threatening the future survival of the organisation. They quote the example of the management failing to respond to market change at Encyclopedia Britannica. The company's management disregarded the threat of CD-ROM technology and the advent of the Internet. As a consequence this 200-year-old company was totally unprepared for the impact of accessing information via free online downloads and faced a huge loss in sales and profitability, which eventually led to the company being sold at a fraction of its historic value.

As summarised in Figure 7.1, Cravens et al suggest consideration of a number of variables which will assist in defining the most appropriate future for the organisation. Three of the factors of influence are concerned

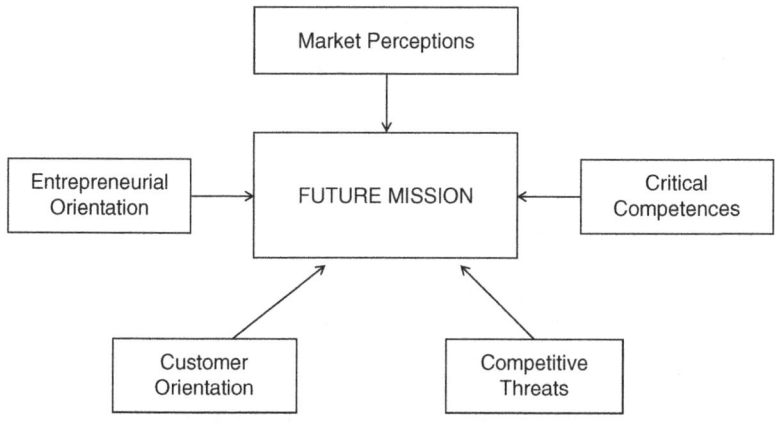

Figure 7.1 Mission and factors of influence

with external variables. One is the management's perceptions about the probable nature of future markets including when, how and why these are expected to undergo change. Linked to this assessment is how the organisation's orientation towards meeting the needs of customers will be influenced by changes in market conditions. Few organisations operate in isolation. As markets change it can be expected that resulting new strategies by competitors will need to be countered by the organisation.

An organisation usually should consider establishing a culture based upon exhibiting an entrepreneurial orientation and be capable of defining which competences will be critical in supporting this orientation. This orientation is of most benefit when an organisation's existing critical competences fail to provide the basis for responding to perceived market change and is particularly important when competences need to be re-directed towards exploiting a range of very different customer and competitive situations.

Companies that achieve superior performance on the basis of an entrepreneurial orientation tend to exhibit the characteristics of continuous learning. This activity leads to ongoing re-assessment of the perceived nature of future markets and whether further refinement of mission is a necessary action. One of the greatest threats facing organisations in a recession or an extended period of reduced customer spending is 'value migration'. This is when customers, who due to economic circumstances have been forced to reconsider their buying behaviour, no longer buy existing products or services but migrate to new propositions which are perceived as more adequately meeting their value requirements. To avoid the threat of value migration the organisation must be able to sense change, comprehend how customers might alter their buying habits and the probable actions that competitors might implement to retain or grow their business (Day, 1994).

One possible approach to assist deliberations over how to respond to a prolonged period of austerity is to construct a customer product space map of the type shown in Figure 7.2. In this diagram the proposal is that there exist four different purchasing scenarios each exhibiting the following features:

(1) *Reasoned purchasing*, involving necessity goods which customers have little choice but to continue purchasing. The purchase rationale is often accompanied by an acceptance that obtaining the benefit of higher performance or quality will necessitate paying a higher price.
(2) *Budget purchasing*, involving necessity goods which customers have little choice but to continue to purchase. The purchase rationale is that decisions will be determined by affordability or price.
(3) *Consumptionism purchasing*, involving optional goods which customers can choose whether or not to purchase. Where the customer does decide they are in a position to buy such goods their purchase rationale is often biased towards achieving the benefit of higher performance or quality, accompanied by an acceptance of the need to pay a higher price.
(4) *Impulse purchasing*, involving optional goods which customers can choose whether or not to purchase. Where the customer does decide they are in a position to buy such goods their purchase rationale is usually determined by affordability or price.

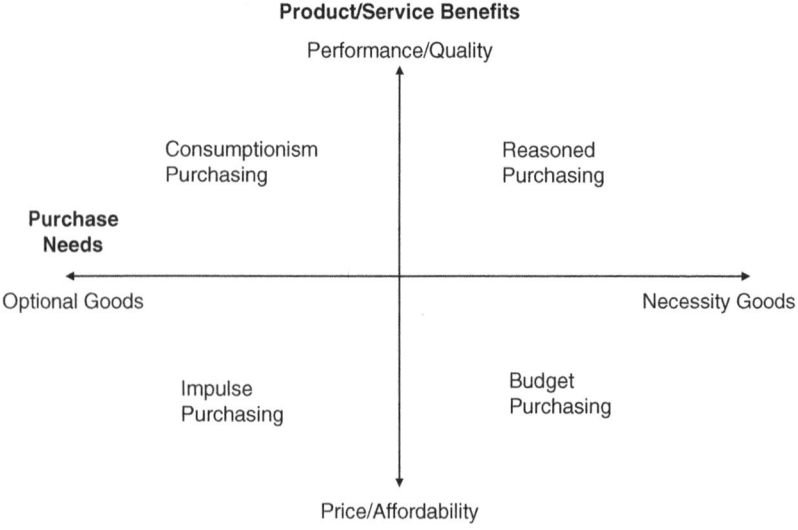

Figure 7.2 Customer product space map

In relation to customer behaviour during a long period of austerity, organisations which supply necessity goods are less likely to be faced with a significant sales downturn. Nevertheless these organisations do need to determine whether any behaviour shift can be expected among current customers in terms of some leaving the market or reconsidering goods being classified as optional instead of remaining a necessity. Even in those cases where future sales might be impacted by either of these behaviour shifts, necessity goods suppliers are in a much better position than companies marketing optional products or services.

Strategy

Although vision provides a definition of purpose, organisations also need to determine the process whereby performance objectives are achieved. This outcome is usually defined by an organisation's strategy statement. Few organisations can expect to be insulated from the activities of competitors. Hence most strategies will require a specification of the 'competitive advantage' to be utilised in the battle to outperform the other organisations also attempting to serve the same customer(s) and market sector(s).

Few organisations can expect to enjoy the luxury of a competitive advantage that can remain unchanged over a prolonged period of time. A more usual scenario is that in order to remain successful organisations must retain flexibility in their choice of competitive advantage (Markides, 2004). This is necessary for the organisation to retain an appropriate fit between market needs and the products or services supplied. One of the benefits of being an entrepreneurial organisation is that up-dating or revising competitive advantage is an accepted and well-understood key managerial philosophy (Chaston, 2011).

The perspective that organisations have the option of transactional versus relationship marketing will influence organisational strategy. The other key dimension of choice is whether the organisation will seek to achieve a competitive advantage through exploiting innovation to offer superior products or services, or seek to maximise the affordability of goods that are to be made available to the market. By combining these two dimensions Chaston (2009) proposed that organisations seeking to exploit an entrepreneurial strategy to respond to the new austerity are able to select from the four alternative options summarised in Figure 7.3.

The organisational attributes associated with entrepreneurial strategy options summarised in Figure 7.3 are:

(1) *High Affordability Strategy*
    (i) Product/services prices or affordability significantly lower than the rest of market.
    (ii) Skilled in the production of 'no frills' products or services.

116 *Entrepreneurship and Innovation During Austerity*

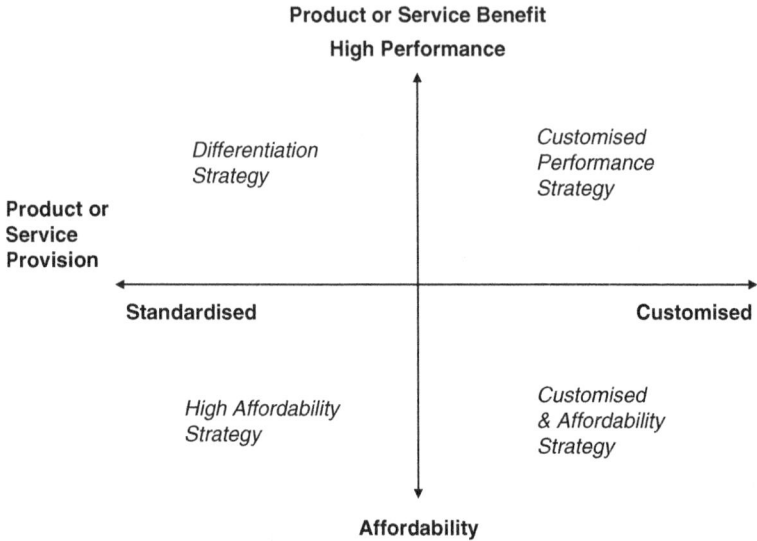

Figure 7.3 Entrepreneurial strategy options

    (iii) Excellence in acquiring prior generation technology and capital equipment at either zero or low cost.
    (iv) Information system designed to rapidly identify adverse cost variance trends across the areas of procurement, production and distribution.
    (v) Culture of employees always striving to find ways of applying conventional thinking to further reducing operating and/or overhead costs.
(2) *Differentiation Strategy*
    (i) Standardised product/service propositions offering outstanding superior performance versus competition.
    (ii) Orientation towards always seeking to extend the performance boundaries of existing products/services.
    (iii) Excellence across the entire workforce in understanding how the latest advances in technology might be incorporated into products/services and/or production processes.
    (iv) Culture of employees always striving to apply conventional approaches to finding new market opportunities for exploiting identified product/service performance improvements.
(3) *Customised Performance Strategy*
    (i) Product/service combination which delivers complete customer specific solutions.
    (ii) Product/services solution based on specifications appropriate for specific customers in market sectors in which an organisation is located.

(iii) Employee obsession over finding even more effective solutions to specific customer problems.
(iv) Information systems which can rapidly identify errors in solution provision.
(v) Culture of all employees committed to working closely with counterparts within customers' organisations.

(4) *Customised Affordability Strategy*
(i) Product/service combination which delivers customer specific solutions that fulfil customer affordability/value/price needs.
(ii) Affordability based on specifications appropriate for specific customers in market sectors in which an organisation is located.
(iii) Employee obsession over finding even more affordable solutions to specific customer problems.
(iv) Information systems which can rapidly identify errors in solution provision.
(v) Culture of all employees committed to working closely with counterparts within customer organisations.

A critical objective of a selected strategy is to ensure a company's long-term survival. There is the risk that a poorly defined strategy may only provide a temporary market advantage, which might dissipate as the level of economic austerity deepens or competitors launch more innovative propositions. This view is supported by Mintzberg (1994, p. 15) who concluded, in relation to strategic planning, that there is 'the danger that the strategy will be outdated within 3 months. If you go back to that strategy you may be focusing your attention on the wrong areas in the business'.

To ensure strategy sustainability, one of the key decisions must be to determine the focus of future innovation. One typology which can be utilised to examine this issue is that proposed by Chaston (2011), who suggested there are two dimensions influencing innovation; namely focus on internal processes versus outputs and provision of new outputs versus sustaining the delivery of existing outputs. Both dimensions can be treated as continuums. This approach generates the typology illustrated in Figure 7.4 which permits identification of these four different potential areas of focus:

(1) *Efficiency innovation*, which focuses upon innovation in relation to the internal processes that can enhance the future delivery of existing products or services.
(2) *Effectiveness innovation*, which focuses on innovation in relation to enhancing the provision of output for existing products or services.
(3) *Development innovation*, which focuses on innovation that utilises changes in internal process which can support the development of new products or services.

118 *Entrepreneurship and Innovation During Austerity*

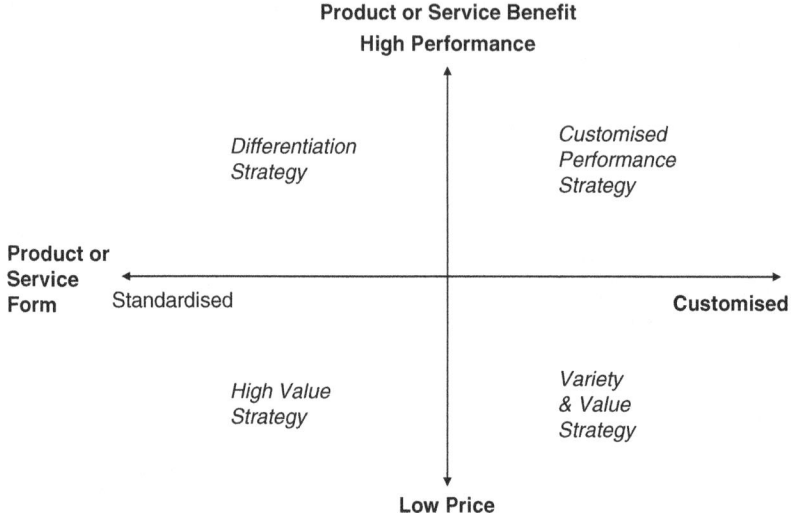

Figure 7.4 Strategic positioning options

(4) **Benefit innovation**, which focuses upon innovation directed towards the creation of new services that enhance the organisation's future products or service provision portfolio.

Flexibility

The onset of the new age of austerity means that external environments have become more unstable and this instability can be expected to increase over time. Intensity of competition will be exacerbated as new entrants from the emerging nations enter Western markets offering lower priced goods. In the face of more difficult trading conditions, organisations must exhibit a high degree of strategic flexibility in order to survive. This will involve exploitation of entrepreneurship and innovation to identify and implement new strategies in response to the worsening economic conditions that exist during a lengthy economic downturn.

In reviewing how organisations can sustain their performance in the face of adverse external environmental conditions, Hitt et al (1998, p. 24) posited that 'Strategic flexibility, then, is the capability of the organisation to pro-act or respond quickly to changing competitive conditions and thereby develop and/or maintain competitive advantage'. In their view, strategic flexibility requires existence of a leadership team that has the vision to comprehend how the organisation can best exploit new opportunities and threats as they emerge in the external environment. The organisation's leadership team must be able to identify ways of managing existing operations to sustain cash flow while concurrently developing new product or service opportunities which will generate revenue. This is necessary because customer behaviour shifts that will emerge during a long period of austerity can be expected to result in declining sales for existing goods.

Further understanding of strategic flexibility was generated by Miles et al's research in the 1970s. They identified a strategic typology whereby organisations could be classified as prospectors, defenders, analysers and reactors (Miles et al, 1978). The researchers proposed that the changing nature of external environments have caused organisations at certain points in their life history to confront the need to reconsider future strategy and internal operations. This activity was labelled as an 'adaptive cycle' in which the organisation examines the need to adopt a more entrepreneurial strategy and to make revisions to the two internal domains of process technology and administrative systems in relation to key issues such as structure, departmental roles and employee skills.

Within the Miles et al's typology prospectors are those organisations that are orientated towards identifying new external environmental and product/service opportunities. The organisation has sufficient flexibility in relation to technology and administrative processes to implement appropriate actions to exploit an identified new opportunity. Defenders focus on retaining control over their current business and attempt to sustain financial viability by exploiting high levels of efficiency in managing current technology and administrative processes. To avoid encounters with other potential suppliers, defenders usually attempt to occupy a small, narrow market domain where their expertise permits delivery of a superior product or service offering little risk of threats from competition. Reactors are the least responsive type of organisation. Management is fixated on sustaining the organisation's existing strategy and internal processes. The usual outcome is that as external environments change this rigidity eventually leads to the demise of the organisation. The fourth strategic typology, which lies halfway between prospector and defender, is the analyser. These organisations monitor environmental trends but will not consider development of new products/services or moves into new areas of customer provision until there is solid evidence that real opportunities exist. Once the organisation is convinced of the need for entrepreneurial action, adequacy in relation to internal flexibility permits an effective re-allocation of resources to support revisions in technology and administrative processes.

In recent years there has been declining academic interest in utilisation of this typology. One reason has been researchers who encountered difficulties in reaching definite conclusions about the relationships which exist between the typology and the performance of organisations in different sectors. Another issue has been the variation between approaches used to classify organisations and the application of scales to achieve an empirical basis for determining which specific firms fit into the four different organisational types (Zahra and Pearce, 1990). Another problem has been the tendency of some researchers to examine the nature of the environments confronting organisations in a specific sector without engaging in the longitudinal data acquisition process needed to identify how the performance of organisations within a sector has changed over time.

Sanchez (1997) proposed that resource flexibility will be greater in those cases where a resource can be applied to a large range of alternatives. Further enhancement in resource flexibility will occur when switching from one use of a resource to an alternative use is a relatively simple activity and does not incur high costs. Another influencing factor is the time required to move a resource from one application to another inside the organisation. The level of strategic flexibility will probably be reduced where a resource to be transferred is in short supply. Although this situation could be avoided through the ownership of slack resource capacity, this solution can be quite costly. The alternative to retaining excess levels of a key resource is to wait until an early indicator of the need for future strategic change has been identified. At this juncture the organisation can assess the implications of change and begin to invest in the acquisition of the additional resources which will soon be required.

Embedding a philosophy of strategic flexibility into an organisation requires considering which strategic planning approach is appropriate for achieving this aim. A 'deliberate strategy' is one which is carefully planned and then implemented. An 'emergent strategy' is a strategy developed over time through a series of activities aimed at finding ways of improving future performance. The deliberate approach runs the risk of being excessively rigid and mechanistic, whereas the emergent process is both informal and flexible. The latter also offers the advantage of involving employees at all levels of the organisation with everybody gaining further understanding of changing external opportunities and threats facing the business.

Increasing environmental complexity makes making sense of futures more difficult and adaptation to change environment more problematic. Mason (2007) proposed that increasing levels of environmental turbulence have reduced orderly competition, thereby increasing the difficulty in accurately predicting customer, product and service requirements. Organisations are faced with a growing need for information that can support strategies focused upon innovation and faster new product development. To survive organisations need to be increasingly vigilant and develop greater capability to rapidly respond to environmental change. An orientation known as 'strategic complexity theory' (Cuhna and Da Cunha, 2006) has been identified. This perspective posits that organisations are complex adaptive systems that, to be successful, must learn to align their strategies with their rapidly changing external environments utilising interaction and response rather than analysis and planning.

Child and McGrath (2001) suggested that adopting a complex orientation means simplicity becomes a major feature of new organisational forms. This is achieved by permitting organisational operations located near to market to adopt simpler structures and granting them semi-autonomous control over their strategic response to change. Structural simplicity is an important facilitator of rapid response because employees are free to tackle problems at the local level, not having to wait for actions by others elsewhere within the organisation. The role of head office is to act as a co-ordinator. This permits retention of an overall common purpose and ensures that learning and knowledge transfer occurs across the entire organisation.

Blue ocean strategy

Case Aims: To illustrate how companies can avoid expensive, non-productive brand wars through exploitation of attribute-based entrepreneurial innovation.

Marketers have long understood that once a product or service moves into maturity or decline on the product life cycle (PLC) curve the usual strategy for sustaining growth is to attempt to steal sales from competitors. A popular strategy for achieving this goal is a head-to-head confrontation in which companies use techniques such as aggressive promotional spending or deep price cuts in an attempt to stimulate customers switching behaviour. In most cases the winner in these conflicts is the organisation that is able to commit the greatest level of financial resources to the battle. In their analysis of the influence of alternative strategies on long-term performance, Kim and Mauborgne (2005) labelled this type of market response strategy as a 'Red Ocean' event. From their analysis of over 150 company case histories they concluded that Red Ocean events typically result in little or no revenue growth, increasing commoditisation of goods and decreased customer loyalty.

Given that most products or services are located on the mature or decline phase of their PLC curve, it can be expected that during the new austerity in the face of shrinking total customer expenditure the number of Red Ocean events will increase dramatically. In seeking to identify a more productive strategy to sustain business growth, Kim and Mauborgne proposed that companies should perceive themselves as explorers and adopt the philosophy of seeking to discover new forms of customer demand. Their recommended approach for achieving this aim, which they labelled a 'Blue Ocean' strategy, is for companies to redefine the benefit offered by their product or service as the basis for occupying market niches where there is little or no competition.

The basis of the Blue Ocean philosophy is that customers make purchase decisions based upon the attributes, availability and price of goods. Their recommended approach is for managers to exhibit an entrepreneurial orientation and seek to develop innovative bundles of attributes which break accepted cost-differentiation trade-offs within a market. To assist in the execution of a Blue Ocean approach Kim and Mauborgne proposed that, in seeking to redefine benefits being offered to customers, managers should assess how this can be achieved by eliminating, reducing, raising or creating new attributes being offered by a product or service.

Kim and Mauborgne presented the example of the cuddly toy animal market. This market is constituted of manufacturers (i) competing on the basis of price at the bottom end of the market and (ii) companies such as Gund which operate at the middle to upper end offering higher quality goods based upon attributes like plushness or visual cuteness. In their view, this is clearly a Red Ocean market because any action to increase sales is typically based upon major promotional spending wars or heavy price cutting. The authors suggested the first attempt to implement a Blue Ocean Strategy within the plush toy sector was the Build-A-Bear Workshop proposition. The Build-A-Bear concept offered the new, unique benefit of allowing customers to experience the joy of creating their own bear. Product pricing for a Build-A-Bear item ranges from $60–100 per bear depending on the number and type of accessories purchased by the customer. The concept supported the creation of a new retail chain through which over 50 million bears were sold in the first 10 years following market launch.

References

Calfee, D.L. (1993), Get your mission statement working!, *Management Review*, Vol. 82, No. 1, pp. 54–59.
Chaston, I. (2009), *Entrepreneurship and Small Firms*, Sage, London.
Chaston, I. (2011), *Public Sector Management: Mission Impossible?*, Palgrave Macmillan, Basingstoke.
Child, J. and McGrath, R.G. (2001), Organizations unfettered: organizational form in an information-intensive economy, *Academy of Management Journal*, Vol. 44, pp. 1135–1148.
Collins, J.C, and Porras, J.I. (1995), Building a visionary company, *California Management Review*, Vol. 37, No. 2, pp. 80–102.
Collins, J.C. and Porras, J.I. (1996), *Built to Last*, Century Business, London.
Cravens, D.C., Piercy, N.F. and Prentice, A. (2000), Developing market-driven product strategies, *The Journal of Product and Brand Management*, Vol. 9, No. 6, pp. 369–388.
Cunha, P.M. and da Cunha, J.V. (2006), Towards a complexity theory of strategy, *Management Decision*, Vol. 44, No. 7, pp. 839–850.
Day, G.S. (1994), Capabilities of market-driven organizations, *Journal of Marketing*, October, pp. 37–52.
D'Intino, R.S., Boyles, T., Neck, C.P. and Hall, J.R. (2008), Visionary entrepreneurial leadership in the aircraft industry: the Boeing Company legacy, *Journal of Management History*, Vol. 14, No. 1, pp. 39–54.

Hickman, C.R. (1992), *Mind of a Manager, Soul of a Leader*, John Wiley & Sons, New York.
Hitt, M.A., Keats, B.K. and DeMarie, S.M. (1998), Navigating in the new competitive landscape: Building strategic flexibility and competitive advantage in the 21st century, *The Academy of Management Executive*, Vol. 12, No. 4, pp. 22–42.
Kantabutra, S. (2008), What do we know about vision?, *Journal of Applied Business Research*, Vol. 24, No. 2, pp. 127–138.
Kantabutra, S. and Avery, G.C. (2010), The power of vision: statements that resonate, *The Journal of Business Strategy*, Vol. 31, No. 1, pp. 37–45.
Kelly, N. (2010), Corning's promising and very profitable future, *American Ceramic Society Bulletin*, Vol. 87, No. 2, pp. 32–39.
Kim, C.W. and Mauborgne, R. (2005) *Blue Ocean Strategy: How to Create Uncontested Market Space and Make the Competition Irrelevant*, Harvard Business School Press, Boston, MA.
Longstaffe, C. (2005), Winston Churchill, a leader from history or an inspiration for the future? *Industrial and Commercial Training*, Vol. 37, No. 2/3, pp. 80–83.
Markides, C. (2004), What is strategy and how do you know if you have one, *Business Strategy Review*, Vol. 15, No. 2, pp. 5–12.
Mason, R.B. (2007), The external environment's effect on management and strategy: A complexity theory approach, *Management Decision*, Vol. 45, No. 1, pp. 10–28.
Miles, R.E., Snow, C.C., Meyer, A.J. and Coleman H.J. (1978), Organizational strategy, structure, and process, *The Academy of Management Review*, Vol. 3, No. 3, pp. 546–562.
Mintzberg, H. (1994), Rethinking strategic planning part I: pitfalls and fallacies, *Long Range Planning*, Vol. 27, pp. 12–21.
Morgan, G. (1986), *Images of Organisation*, Sage Publications, London.
Morris, J.A., Brotheridge, C.M. and Urbanski, J.C. (2005), Bringing humility to leadership: antecedents and consequences of leader humility, *Human Relations*, Vol. 58, No. 10, pp. 1323–1350.
Rowe, G. and Nejad, M.S. (2009), Strategic leadership: short-term stability and long-term viability, *Ivey Business Journal*, Vol. 11, pp. 31–48.
Sanchez, R. (1997), Preparing for an uncertain future: managing organizations for strategic flexibility, *International Studies of Management & Organization*, Vol. 27, No. 2, pp. 71–94.
Shelton, K. (1999), Great leaders, *Executive Excellence*, Vol. 16, No.4, pp. 14–18.
Serling, R.J. (1992), *Legend and Legacy: The Story of Boeing and its People*, St Martin's Press, New York.
Strong, C (1997), The question we continue to ask: How do organisations define their mission?, *Journal of Marketing Practice*, Vol. 8, No. 34, pp. 268–283.
Tischler, L. (2004), Seven traits of great leaders, *Fast Company*, Boston, November, pp. 112–113.
Westley, F. and Mintzberg, H. (1989), Visionary leadership and strategic management, *Strategic Management Journal*, Vol. 10, pp. 17–32.
Zahra, S.A. and Pearce, J.A. (1990), Research evidence on the Miles-Snow typology, *Journal of Management*, Vol. 16, No. 4, pp. 751–767.

8
Innovation Strategies

Classifying innovation

Many brands in the maturity phase of the product life cycle (PLC) utilise a strategy to extend the life of a product by engaging minor product reformulations. These changes rarely affect basic attributes or the primary benefit being offered. This approach is simple to implement and involves minimal costs or risk. Hence the impact on sales tends to be relatively minor although the company can engage in making a promotional claim such as 'new, improved'. This can be contrasted with entrepreneurial strategies where the result is the creation of a product or service which is very different from existing goods, capable of offering the customer a superior benefit promise.

In terms of when entrepreneurship is most important to organisations, Schumpeter posited that the activity was critical to firms seeking to survive cataclysmic events such as the Great Depression of the 1930s. A more recent analysis of the performance of firms during an economic downturn was undertaken by Roberts (2003). He used the Profit in Marketing Strategy (PIMS) database which contains information on over 4,000 businesses from a wide range of industries, located mainly in North America and Europe. Roberts concluded that those companies which survived major recessions and were able to exploit the benefits of a leadership position once economic conditions begin to recover are those which have invested in the three key activities of marketing, innovation and delivering superior customer quality.

Gartner (1988) argued that entrepreneurship can be defined in terms of innovative behaviour that is allied to a strategic orientation of pursuing business growth and profitability. Miller (1983) proposed firms which exhibit an entrepreneurial orientation will be staffed by top managers who are willing to take risks, favour change and seek to exploit innovation in order to obtain a proactive competitive advantage over other firms. There have been a number of empirically-based efforts to define the entrepreneurship in terms of personality traits, attitudes and behaviours of business leaders. Stewart et al (1998) found that entrepreneurs exhibit higher ratings in relation to a desire for achievement, a risk-taking propensity and a strong preference for involvement in innovation. Hyrsky (2000) identified a strong work ethos, energy, innovativeness, risk taking, ambition, achievement and egotistic features as dimensions of entrepreneurship. Georgelli et al (2000) described 'being entrepreneurial' as having a willingness to take risks, being innovative and having an ambition to grow the business. They suggested that the core competences for entrepreneurship are a capacity for changing business processes, launching new products and services and effective planning.

Covin and Slevin (1988) defined entrepreneurial style in terms of the extent to which managers are inclined to take business-related risks, favour change, are innovative and compete aggressively with other firms. Hamel and Prahalad (1994), commenting upon innovation in the large firm sector, suggested that entrepreneurial behaviour often leads to the emergence of a completely new benefit proposition. These researchers used examples of observed significant change in various industrial sectors to propose that the influence of unsatisfied market need frequently causes entrepreneurial firms to break with convention and exploit an emerging opportunity through the provision of a new, more innovative solution. This perspective was also reflected in Chaston (2000, p. 21) who suggested that entrepreneurship could be defined as 'the behaviour exhibited by an individual and/or organisation which adopts a philosophy of challenging established market conventions during the process of developing new solutions'. Hamel and Prahalad further proposed that major changes in industrial sectors have occurred because a company has been prepared to change the rules of the game. In their view, to take control of the future a company must (i) change in some fundamental way the rules of engagement, (ii) redraw the boundaries between industries and/or (iii) create entirely new industries.

One approach to determining the scale of innovation associated with the observed strategies of organisations is to classify outcomes in relation to the dimensions of (i) level of breaking with existing market conventions and (ii) the degree to which the benefit proposition has been changed. As summarised in Figure 8.1, this taxonomy generates four possible outcomes. In the case of 'New Improved Innovation' the product or service benefit remains unchanged and no market conventions are altered by the offered goods. Similarly in the case of 'value innovation' no market conventions are changed but the market is offered a new benefit proposition. In the case of the other two alternatives proposed in Figure 8.1, innovation involves challenging existing conventions. With 'disruptive innovation' this is accompanied by customers being offered an unchanging benefit, whereas in the case of 'radical innovation' the new benefit is very different to that previously available to the market.

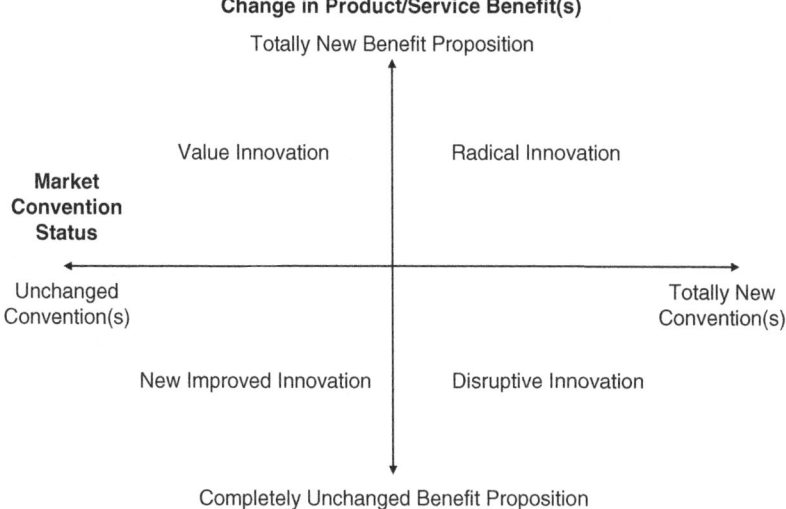

Figure 8.1 Alternative innovation propositions

Succeeding at being innovative

Case Aims: To illustrate the issues associated with overcoming organisational obstacles and enhancing the potential for success from involvement in innovation.

Given that survival during the new austerity is likely to be determined by organisations being more innovative, the issue arises of how this can be achieved. Worryingly even before the onset of the global recession, research by the consultancy firm Strategos involving managers in over 550 large companies revealed that the majority were critical of their organisations' abilities in this crucial area of activity (Loewe and Dominiquini, 2006). The top obstacles to innovation identified by respondents were:

(1) Excessive focus on immediate, short-term performance.
(2) Inadequate allocation of resources or staff.
(3) A tendency of senior managers to expect fast payoffs from projects.
(4) The absence of systems and structures to effectively implement the innovation process.
(5) A strong belief within the organisation that innovation is an inherently risky activity.

To overcome the obstacles, Loewe and Dominiquini proposed that the management of following elements are critical:

(1) *Leadership,* in terms of senior management demonstrating to the entire workforce that their personal commitment to innovation is the most important driving force in determining priorities, policies and the allocation of scarce resources within their organisation.
(2) *Processes and systems,* which ensure creativity receives support, new ideas are generated, ideas are converted into viable propositions and staff have sufficient autonomy to access required resources.
(3) *People and skills,* reflecting the need to harness innovation talent throughout organisations such that a diversity of knowledge, skills and competences can be fully exploited.
(4) *Culture and values,* focusing upon creating an organisational environment where innovation is valued, people are permitted to make mistakes and the occurrence of errors is perceived as a critical dimension of learning.

Strategos promote the idea to their clients of engaging in 'innovation diagnostics' to identify an organisation's innovation enablers and the obstacles confronting effective innovation. The company's experience suggests the diagnostic should include (i) a review of the opportunity pipeline and market place and (ii) a 'health check' of innovation capability in relation to leadership, processes, people and culture. To determine the potential offered by a new product, Strategos recommend a review of the company's innovation track record to assess whether there exists a healthy mix of incremental and breakthrough innovations. In terms of ensuring the probability of future success, an organisation will need to examine:

(1) What customers are being served and whether new markets could be opened or if new or untapped segments exist.
(2) What new or improved benefits could be made available.
(3) What should be the future position in industry supply and whether other partners can assist in more effective serving of customer needs.
(4) What level of revenue can be generated from innovation and whether the associated cost/benefits are sufficient to reach defined profitability goals.
(5) Whether opportunities exist to protect new innovations from competitive threats and how long these innovations can be protected from such threats.
(6) What competences exist or can be developed that offer superiority over existing and future potential competition.

Value innovation

Kim and Mauborgne (1999) concluded that many firms perform poorly because of a strategic orientation focused on a reactive response involving 'me too' propositions and a failure to understand the real nature of customer need. They recommended utilisation of 'value innovation'. This involves offering new and superior value to customers in existing markets or alternatively, establishing new market segments. To achieve this aim they suggest that firms need to:

(1) Challenge the apparent inevitability of existing industry sector conventions.
(2) Seek to identify ways in which the value offered to the customer can be dramatically improved.
(3) Be prepared to start over when value enhancement activities are not feasible due to weaknesses in existing organisational competences.
(4) Focus upon customer value solutions that significantly exceed those which are perceived as acceptable by other organisations in the same industrial sector.

Dillon et al (2005) concluded that many organisations over-emphasise technology as a source of innovation when in fact superior financial performance can be achieved using value innovation, which can occur with little or no expenditure on new technology. These authors noted that value innovators are often not the first market entrants but succeed by creating customer demand by offering a leap in value at an accessible price. The activity must be based upon authentic customer demand that will respond positively to a new idea. Ability and effectiveness of an organisation to be a value innovator is strongly influenced by internal culture in which trust, integrity and honesty are fundamental behavioural traits.

Dillon et al developed a 'Value IQ Instrument' which can be used to assess key aspects of the value innovation process. Some of the factors the instrument assesses include (i) organisational learning processes, (ii) breakthrough options, (iii) degree of external focus and (iv) whether the value chain is being fully exploited. These authors have also proposed that management of the value innovation process (or VIP) involves the following five phases:

(1) *Phase 1 Business Intelligence*: involving idea generation by drawing upon multiple sources of information seeking whereever possible to exploit the organisation's existing core competences.
(2) *Phase 2 Value Modelling*: involving validating the reality of the value innovation by relying upon inputs from key customers.

(3) *Stage 3 Prioritisation*: using market research to validate the value proposition is perceived as superior by customers and offers the company an adequate cost/benefit.
(4) *Stage 4 Implementation*: involving assembling the project team to manage the launch of the new proposition.
(5) *Stage 5 Final Validation*: assessing customer superiority and whether cost/benefits previously identified will actually be delivered.

Aiman-Smith et al's (2005) study of value innovation identified the importance of both staff within the organisation and other members of the value chain being open to new ideas and being prepared to take risks. A strong customer orientation ensures the successful identification of new products or services, new business models and new markets. These researchers concluded that agile decision-making empowers employees to act without waiting for formal approval from senior management. Finding new value opportunities demands the existence of a business intelligence system capable of detecting market trends and understanding the competition. This system is only truly effective where there is open communication throughout the organisation with the system being used to support activities such as asking 'what if' questions whilst developing new or improved value propositions.

Adding value at Alcan

Case Aims: To illustrate how a company can engage with others in seeking value innovation opportunities.

Goodrich and Aiman-Smith (2007) examined value innovation within Alcan Global Pharmaceutical Packaging. The company recognised that in an increasingly competitive market where there is pressure to find ways of reducing healthcare delivery costs, survival demanded new ways of working with customers. The stated aim was to find ways of delivering exceptional value to key customers.

The project start point was to determine what customers perceive to be the role of Alcan products in supporting customers' needs, in order to use packaging to market their products and to protect these products as they pass through the healthcare distribution chain. This knowledge was acquired through internal and external exploratory research. To learn from customers Alcan engaged in productive dialogue to avoid being perceived as seeking to sell products. These dialogues were used to learn more about organisations further downstream in the healthcare value chain. Alcan staff concurrently engaged in dialogue with major stakeholders such as medical insurance companies and public sector procurement agencies about future problems in the healthcare sector. Acquired knowledge was used to develop a specification of the potential wants and needs of key customer groups. Where this specification required further knowledge before a solution could be evolved, further research was undertaken. Having acquired adequate understanding, staff sought to identify solutions which could be technology, process or service based. The final phase was to review potential solutions with customers and, where necessary, gain their involvement in 'co-creation'. The first outcomes from the company's engagement in value innovation were products to counter drug counterfeiting and the development of new packaging that supports actual delivery of a medical treatment.

Other value innovation approaches

Sheehan and Vaidyanathan (2009) suggested that drawing upon accumulated experience and knowledge provides the basis for offering greater value through innovation. One approach is 'industrial efficiency logic', which involves finding new ways of making industrial processes more productive in order to produce higher volumes of standardised output. This permits goods to be offered at a lower price. An example of this approach is Huish Detergents, the largest manufacturer of private label dish and laundry detergents in North America. The company invested in internal business process improvements and developed a highly integrated supply chain computer-based information system to offer quality comparable with national brands but at significantly lower prices.

Another strategy is 'network services logic', involving the creation of supply networks that link together customers and suppliers. Participants can exploit information and knowledge interchange to identify new opportunities. A key aim in such networks is to add value. The simplest approach is to expand use of automated pre-purchasing support systems and enhance after-sales services. The network can also be utilised to permit members to link into each others' customer base. An example is provided by J Boats, Inc. which sells premium priced yachts to racing enthusiasts. The company increased their product offering by permitting customers to become members of J Boat's racing class associations, which organise regattas for owners to race their J Boat sailboats. Through these associations members can also exchange ideas on how the company's boat designs can be enhanced to improve performance.

Another value innovation opportunity is 'knowledge intensive logic', involving the supplier utilising the knowledge acquired from analysis of buyer behaviour in order to develop customised products or services. This approach can be enhanced by the supplier developing systems whereby the customer can actually participate in the customisation process at the pre-purchase phase; thereby allowing the creation of a proposition totally designed to meet their own specific personal needs. An example of a company which effectively exploits this form of value innovation is Dell Computers.

> ### B2B value innovation
>
> *Case Aims: To illustrate how value innovation can protect an organisation from major changes in market conditions.*
>
> The advent of the Internet put many distributors at risk because customers are able to link directly with suppliers and no longer need the services of a distributor. The process is known as 'disintermediation'. Marshall Industries is one of the largest B2B distributors in the US electronics component industry. The senior management were very alert to the threat posed by the Internet and determined that to retain customers it was necessary to exceed customers' expectation. Marshall articulated the new benefit proposition of 'free, perfect and now'. Absolute provision of this benefit is impossible, but the company sought to approach complete achievement by restructuring their operation to support the needs of an Internet mediated world (El Sawy et al, 1999).
>
> In terms of 'free', Marshall used value innovation to add incremental value to products, causing customers to perceive financial advantages in sourcing their products from the distributor and not buying direct from manufacturers. The company invested in the creation of a computer-based system that provided an internal dictionary to assist in answering customer questions, as well as an online design service and an online educational programme to help customers learn about the latest advances in the IT industry and how these would influence the performance of the next generation of electronic products.
>
> In relation to 'perfect', and to avoid market commoditisation, Marshall developed the capability to customise products to suit the specific design and manufacturing needs of individual customers. The company undertook research to identify emerging advances in new technology and advised customers on how these might lead to the availability of superior next generation components. Customers could be confident of being able to purchase goods offering the latest technology and knowing what next generation components should be specified when developing new products based upon exploiting leading edge technology.
>
> The benefit of 'now' essentially sought to meet customers' desire for instant deliveries. Marshall developed systems for more effectively

managing of inventories and used computer-based information interchange with customers to gain an in-depth understanding of customers' purchase patterns. Where instant delivery was deemed critical Marshall revised their distribution systems to include holding key components instore, either near to or actually located inside customers' premises. The company recognised that value innovation demanded revisions to structure and employee management practices and so moved from a hierarchical to a flatter organisational structure, with employees granted much greater power and authority to make decisions without seeking prior approval from their line managers. The departmental orientation was revised such that support staff now reported directly to a front line client manager to ensure that a proactive, highly responsive interaction system existed for linking the customer to relevant staff inside the company.

Marshall avoided the threat of disintermediation and revenue doubled over a five-year period. Value innovation strategy enhanced employee productivity and doubling the size of the company's operations was achieved with virtually no increase in the total employee numbers.

Disruptive innovation

Christensen (1997) proposed that the orientation of most large corporations is towards focusing their R&D efforts on introducing incremental improvements in existing products or organisational processes. The potential problem with this philosophy is that the future performance of these firms is highly vulnerable to a new player entering the market offering a significantly different product or a more effective production process. The actions of a firm challenging existing industry conventions was labelled by Christensen as 'disruptive innovation'. An example of disruptive innovation is provided in the PC industry when Dell used direct marketing of computers to US consumers whilst existing major computer manufacturers continued to rely on a direct sales force, distributors or retail outlets.

Christensen's original ideas were developed during a period of economic prosperity but the concept remains relevant even in an age of prolonged austerity. This is because disruptive innovation can offer superior value through developing a new business model that offers a more affordable product benefit in existing markets. The current downturn has fundamentally changed customers' needs, suggesting that winning organisations will exploit disruptive innovation to sustain the availability of similar benefits whilst concurrently offering more affordable product or service propositions. Probable losers will be those firms who either try to ride out the recession by merely reducing prices or rely upon very limited innovation-based strategies to launch minor improvements to their existing products or services (Leavy and Sterling, 2009).

Christensen et al (2002) concluded that the probability of creating a successful, new growth business is 10 times greater when innovators pursue a disruptive strategy instead of focusing upon developing a new improved product based upon existing technology. Disruption often delivers the benefit of providing customers with a lower cost proposition. The increasing speed with which technological change occurs in many industries means disruptive innovation has now become much easier to implement. Christensen et al proposed that the probability of success will be enhanced where one or more of the following conditions apply:

(1) *Enhanced affordability*, permitting the innovator to embrace customers who in the past could not afford the cost of accessing the offered benefit. An example of this philosophy is provided by online retail stock broker Charles Schwab whose exploitation of technology permitted private citizens to engage in trading shares without utilising the services of expensive professional services offered by firms such as Merrill Lynch.
(2) *Enhanced simplicity*, making the product or service attractive to people who believe the existing offerings are too complex. An example is NTT DoCoMo, which attracted millions of new subscribers to their wireless Internet-access services by making it easier to download ring tones and online games.
(3) *Enhanced economy business model*, permitting the innovator to make the benefit available at a lower cost; thereby offering the dual opportunity of attracting new users into the market and existing customers to switch from competition. An example of this concept is provided by the economy airlines, where consumers who rarely used airlines began to fly more and business travellers switched from more expensive full service carriers.

Paap and Katz (2004) questioned the conventional theory of a large firm's failure being attributed to a lack of recognition of the scale of the threat posed by a new firm entering their market. However they do support Christensen's view that market leaders' responses to changing market circumstances is often constrained by major customers' insistence on their suppliers concentrating on making improvements to existing products. Thus large firms may be forced to remain focused on product or process innovation which can permit the organisation to stay ahead of other large organisations in the same market sector (Demuth, 2008). Thus, for example, IBM did actually recognise the potential of the mini-computer to provide smaller organisations with access to more affordable computer technology. However demands from the firm's existing customers influenced IBM's decision to continue to develop the next generation of mainframe computers capable of offering even faster, more powerful data processing capability. This behaviour permitted Ken Olsen to launch the Digital Equipment Corporation (DEC), operating with a strategy of making computers affordable by supplying the first generation of mini-computers.

There are a number of well-documented examples of poor organisational performance that validate Christensen's viewpoint. Nevertheless there is also evidence to suggest his theory is not a universal explanation that can be applied to explain inappropriate strategic thinking in all industrial sectors (Cravens et al, 2002). For example, case materials in most branded consumer goods sectors would suggest only a minority of large company performance downturns can be explained by the advent of a disruptive technology. This is because the expertise which exists within these major corporations should permit an immediate response to a newly emerging market threat by utilising their huge internal resources to mount a successful counter attack. For example, when Canon first started to make inroads into the photocopier market by exploiting innovative manufacturing techniques to offer their lower price photocopiers, Xerox clearly had both the technical expertise and dominant market position which would have permitted them to defeat their new enemy. Such examples of market disruption caused by innovation suggest that large organisations would do well to reflect on Parnell et al's (2005) conclusion that organisational leaders 'should resist the notion that to-day's source of competitive advantage will be eternal'.

Radical innovation

Radical innovation is typically the preserve of the entrepreneur, individuals or small organisations who have the creativity, skills and knowledge to develop totally new benefit concepts as a consequence of being willing to challenge existing industry conventions. For the developed nations facing both reduced customer spending and increasing threats from the emerging economies, a very appealing long-term survival strategy is one reliant upon radical innovation. This is because radical innovation can result in sustaining a superiority benefit over competition involving unique, specialist knowledge which is rarely easy to replicate. Furthermore in some cases radical innovation can provide the basis for the creation of highly profitable new industries. Examples of the benefits of the latter scenario are demonstrated by the success of American companies such as Apple, IBM and Google in exploiting information technology as the basis for developing new-to-the-world business propositions.

Many large corporations can be expected to fail to respond to the problems created by the new austerity. This is because they lack the ability to think entrepreneurially and hence are unable to exploit radical innovation to redefine their market position, enter new markets or create new-to-the-world markets. Stringer (2000) concluded there are a number of factors which impair organisations' abilities to engage in radical innovation. One is that senior managers often perceive embracing emerging, non-traditional technologies as costing too much money and that a change in strategy will lead to a decline in the revenue flows from current operations. As a consequence there is a strong preference for making incremental improvements to current products, services and core technologies. Another obstacle is the structure of large companies. Although being big can confer scale advantages these are often accompanied by inflexible, bureaucratic structures that discourage entrepreneurial behaviour. Radical innovation usually requires dramatic shifts in production capabilities, distribution mechanisms or customer relationships which have the potential to threaten the status quo, hierarchies and existing senior managers' power base.

Stringer posited that entrepreneurial behaviour demanded in a time of rapid environmental change can be inhibited by large internal R&D operations. This reflects the research staff's preference to stay with existing technologies and to continue to exploit their years of accumulated knowledge and experience. There is also the issue that to initiate a re-direction in research programmes, the budgets for existing programmes may have to be cut, a move which will be strongly resisted by the teams engaged in these current programmes. The R&D teams' objections will possibly be supported by sales and marketing who are concerned that proposed changes will also adversely impact existing relationships with key customers.

Given the impact on performance that radical innovation can contribute, Stringer has proposed that large organisations need to implement the following actions in order to promote a stronger entrepreneurial orientation:

(1) Make breakthrough innovation a priority.
(2) Hire more creative and innovative people.
(3) Grow informal project laboratories within the organisation and provide people with the freedom to engage in innovation free from bureaucracy and tight controls.
(4) Create 'idea markets' within the organisation.
(5) Experiment with acquisitions, joint ventures and alliances to develop links with external more entrepreneurial individuals and entities.
(6) Engage in corporate venturing in which entrepreneurs are permitted to create new businesses inside the organisation.
(7) Establish a corporate venture capital fund which entrepreneurs from both inside and outside the organisation can use to fund the idea commercialisation phase of new product or service development.

Colarelli et al (2005) proposed that successful innovation demands specific organisational competences, with the most important of these being that of 'discovery'. This involves the creation, recognition, elaboration and articulation of opportunities. Once a radical new idea has been identified the organisation must create an 'incubation capability' to convert the idea into a viable business proposition. Skills needed for incubation are experimentation, market learning, market creation and market testing. These first two competences can be enhanced by an 'acceleration capability'. This involves developing new approaches to project implementation based upon building the appropriate business infrastructures which ensure the radical new product or service is rapidly developed and adopted by customers. Colarelli et al identified the following four business models for managing radical innovation:

(1) *Competency and Readiness*: continuous deepening and strengthening of the organisation's technical capabilities and science base through sensing opportunities ahead of competition.
(2) *Strategy-Driven*: a small number of technology-market domains are identified in new business arenas containing few competitors, which require advanced technology and offer promising major future market opportunities.
(3) *Execution-Driven*: based around independent initiatives inside the company which are allocated adequate resources to exploit a core new idea to provide the basis for new business units.
(4) *Rationality Model*: roles and responsibilities are clearly defined and understood. The core of radical innovation operations is a central R&D facility and the majority of this group's radical innovations projects will be tightly aligned with the firm's current market activities.

Story et al (2009) confirmed the importance of the three competences of discovery, incubation and project acceleration in the car industry. However in their view, another critical competence is 'commercialisation'. This involves understanding the customer, an ability to pick market winners and the critically important ability of having an accurate insight about potential acceptance of a radical innovation. Another dimension of the commercialisation competence is the capability to manage the supply side of the innovation equation. This is critical in order that the final design can actually be based upon components and sub-systems which the company is able to procure when the actual manufacturing phase commences.

Open innovation

The traditional approach to innovation is to retain ownership of proprietary knowledge by adopting a 'closed innovation' philosophy. This trait reflects an unwillingness to collaborate with other organisations, most usually because of concerns over the loss of proprietary knowledge. As markets have become more volatile and technology more complex, some organisations are beginning to adopt an 'open innovation' philosophy designed to overcome some of the limitations associated with closed systems. Success in open innovation not only benefits the firm which has the original new idea, but also those organisations involved in the collaborative processes that lead to the launch of the new product (Chesbrough, 2003).

Chesbrough has suggested there are four types of individuals involved in open innovation, namely:

(1) *Innovation explorers*, who specialise in undertaking discovery-level research in both public sector laboratories and corporate R&D centres.
(2) *Innovation merchants*, who focus on developing a narrow set of technologies that are codified into intellectual property brought to market or sold to other organisations.
(3) *Innovation missionaries*, who seek to advance new technologies but are not really interested in financial gain and instead wish to serve the common good of society.
(4) *Innovation marketers*, who perform the functions of bringing new products to market by exploiting their expertise in marketing management.

Huang et al (2010) posited that open innovation enables an organisation to be more effective in both creating new products and enhancing existing value-added activities. They proposed that the process also helps create value by leveraging many more ideas from a variety of external concepts and allows greater value capture in the utilisation of the firm's existing assets. Chesbrough (2007) noted not all companies apply the same approach to openness. In his view the process can best be described as a continuum ranging from a low to a high degree of openness. Christensen et al (2002) argued that firms manage open innovation in different ways depending on (i) their position in the innovation system, (ii) the stage of product/service maturity and (iii) the scale of the value proposition.

Lichtenthaler (2009) proposed that open innovation is very useful when an organisation is seeking to accelerate market acceptance for a new technology or new market standard. He found that firms which emphasise radical innovation are not always able to develop all the required knowledge internally. Hence there is a need to rely on complementary external sources to support the creation of a commercially viable new proposition. Lazzarotti et al (2010) posited that firms' increase R&D activities will be accompanied by more involvement in forming collaborative links with other organisations. In their view, open innovators often choose an aggressive technology with the aim of becoming a first mover in existing or new markets. Slowinski and Sagal (2010) proposed that the following actions may increase the effectiveness of open innovation:

(1) Incorporate external thinking into the strategic planning process.
(2) Convert planning outcomes into a set of prioritised project briefs.

(3) Utilise a structured process for the make/buy/partner decisions.
(4) Look inside the company first when seeking new ideas.
(5) Treat collaborative idea searches as a mutually beneficial process.
(6) As new data are acquired, use these to update and further refine the project brief.
(7) Establish and maintain alignment with all internal and external relationships.
(8) Use a structured process for planning and negotiations.
(9) Negotiate with a focus on 'Win-Win' outcomes.

Open innovation may be distributed among a larger number of different actors. This approach is often described as a 'boundary spanning activity'. Elmquist et al (2009) noted there are two key dimensions influencing the innovation process; namely the number of partners involved and the internal versus external focus of the innovation programme. The risk-facing firms, especially in high technology sectors, are those companies that rely too heavily on closed innovation and may miss new market opportunities. This is because many new opportunities may fall outside of the organisation's current business activities and technological competence or can only be exploited by working with other organisations (Chesbrough, 2007). To avoid this outcome management need to recognise that the boundary between a firm and the surrounding environment must be porous; thereby enabling the development of a collaborative approach to knowledge exploitation.

In those cases where open innovation involves collaboration with customers, Fang (2008) suggested there are two approaches; namely customers acting as a passive information resource or being actively involved as co-developers. In many consumer goods markets supply chain networks constituted of suppliers, retailers and distributors are highly connected. As a consequence there is a high level of inter-organisational knowledge interchange. This can assist the speed of new product development. However there is a tendency of some downstream members to be somewhat conservative in response to radical ideas. This may have a detrimental impact on the level of product innovativeness. To overcome this obstacle it is probably better for the supplier to engage in open innovation by also partnering with an end user. Knowledge generated through end-user interaction can then be used to persuade intermediaries of the mutual benefits of adopting a more entrepreneurial orientation.

References

Aiman-Smith, L., Goodrich, N., Roberts, D. and Scinta, J. (2005), Assessing your organisation's potential for value innovation, *Research Technology Management*, Vol. 48, No. 2, pp. 37–42.

Chaston, I. (2000), *Entrepreneurial Marketing*, Macmillan, Basingstoke.
Chesbrough, H.W. (2003), The era of open innovation, *Sloan Management Review*, Vol. 44, No. 3, pp. 35–41.
Chesbrough, H.W. (2007), Why companies should have open business models, *Sloan Management Review*, Vol. 48, No. 2, pp. 22–28.
Christensen, C.M. (1997), *The Innovator's Dilemma*, Harvard Business School Press, Harvard, Mass.
Christensen, C., Johnson, M.W. and Rigby, D.K. (2002), Foundations for growth how to identify and build disruptive new businesses, *Sloan Management Review*, Vol. 43, No. 3, pp. 22–31.
Colarelli, G., O'Connor, M. and Ayers, A.D. (2005), Building a radical innovation competency, *Research Technology Management*, Vol. 48, No. 1, pp. 23–31.
Covin, J.G. and Slevin, D.P. (1988), The influence of organisational structure on the utility of an entrepreneurial top management style, *Journal of Management Studies*, Vol. 25, pp. 217–237.
Cravens, D.W., Piercy, N.F. and Low, G.S. (2002), The innovation challenges of proactive cannibalisation and discontinuous technology, *European Business Review*, Vol. 14, No. 4, pp. 257–268.
Demuth, L.G. (2008), A viewpoint on disruptive innovation, *Journal of the American Academy of Business*, Vol. 13, No. 1, pp. 86–94.
Dillon, T.A., Lee, R.K. and Matheson, D. (2005), Value innovation: passport to value creation, *Research Technology Management*, Vol. 48, No. 2, p. 22–36.
El Sawy, O.M., Malhotra, A., Gosain, S. and Young, K.M. (1999), IT-intensive value innovation in the electronic economy: insights from Marshall Industries, *MIS Quarterly*, Vol. 23, No. 3, pp. 301–314.
Elmquist, M., Fredberg, T. and Ollila, S. (2009), Exploring the field of open innovation, *European Journal of Innovation Management*, Vol. 12, No. 3, pp. 326–345.
Fang, E. (2008), Customer participation and trade-off between new product innovativeness and speed to market, *Journal of Marketing*, Vol. 72, pp. 90–104.
Georgelli, Y.P., Joyce, B. and Woods, A. (2000), Entrepreneurial action, innovation, and business performance: the small independent business, *Journal of Small Business and Enterprise Development*, Vol. 7, No. 1, pp. 7–17.
Goodrich, N. and Aiman-Smith, L. (2007), What does your most important customer want?, *Research Technology Management*, Vol. 50, No. 2, pp. 26–35.
Hamel, G. and Prahalad, C.K. (1994), *Competing for the Future*, Harvard Business School Press, Boston, MA.
Huang, T., Wang, W.C., Yun, W., Tseng, C. and Lee, C. (2010), Managing technology transfer in open innovation: the case study in Taiwan, *Modern Applied Science*, Vol. 4, No. 1, pp. 2–11.
Hyrsky, K. (2000), Entrepreneurial metaphors and concepts: an exploratory study, *International Small Business Journal*, Vol. 18, No. 1, pp. 13–34.
Kim, W.C. and Mauborgne, R. (1999), Strategy, value innovation, and the knowledge economy, *Sloan Management Review*, Vol. 40, No. 3, pp. 41–50.
Lazzarotti, V., Manzini, R. and Pellegrini, L. (2010), Open innovation models adopted in practice: an extensive study in Italy, *Business Excellence*, Vol. 14, No. 4, pp. 11–23.
Leavy, B. and Sterling, J. (2009), Think disruptive! How to manage in a new era of innovation, *Strategy & Leadership*, Vol. 38, No. 4, pp. 5–10.
Lichtenthaler, U. (2009), Outbound open innovation and its effect on firm performance: examining environmental influences, *R&D Management*, Vol. 39, No. 4, pp. 317–330.

Loewe, P. and Dominiquini, J. (2006), Overcoming the barriers to effective innovation, *Strategy & Leadership*, Vol. 34, No. 1, pp. 24–31.

Paap, J. and Katz, R. (2004), Anticipating disruptive innovation, *Research Technology Management*, Vol. 47, No. 5, pp. 13–23.

Parnell, J.A., Von Bergen, C.W. and Soper, B. (2005), Profiting from past triumphs and failures: harnessing history for future success, S.A.M. *Advanced Management Journal*, Vol. 70, No. 2, pp. 36–47.

Roberts, K. (2003), What strategic investments should you make during a recession to gain competitive advantage in the recovery? *Strategy & Leadership*, Vol. 31, No. 4, pp. 31–39.

Sheehan, N.T. and Vaidyanathan, G. (2009), Using a value creation compass to discover Blue Oceans, *Strategy & Leadership*, Vol. 37, No. 2, pp. 13–20.

Slowinski, F. and Sagal, M.W. (2010), Good practices in open innovation, *Research Technology Management*, September/October, pp. 38–46.

Stewart, W.H., Watson, W.E., Garland, J.C. and Garland, J.W. (1998), A proclivity for entrepreneurship: a comparison of entrepreneurs, small business owners and corporate managers, *Journal of Business Venturing*, Vol. 14, No. 2, pp. 189–214.

Story, V., Hart, S. and O'Malley, L. (2009), Relational resources and competences for radical product innovation, *Journal of Marketing Management*, Vol. 25, No. 5/6, pp. 461–481.

Stringer, R. (2000), How to manage radical innovation, *California Management Review*, Vol. 42, No. 4, pp. 70–88.

9
Technology Strategies

Importance

Technological change is possibly one of the most critical of the meta-events that can create a future strategic opportunity or threat for any organisation. This is because technology can provide the basis of totally new industries or permit smaller firms to develop a competitive advantage based upon disruptive innovation which is capable of successfully challenging an industry's market leaders. The risk facing the developed nations during the new austerity in technological innovation is assisting developing nations to more rapidly engage in industrialisation and accelerate the speed with which companies from these areas of the world can begin to move towards achieving global market leadership.

Boston Consulting Group's (BCG) analysis of firms exhibiting the highest capability to survive the current global downturn revealed the top 25 companies have relied heavily upon exploiting innovation to sustain performance (McGregor, 2008). All of these companies have cultures that place a high value on employing creative people in both good times and bad. Their focus is based upon developing a sufficiently diverse portfolio of projects so that there is room for the occasional market mistakes. The top 10 companies in the BCG analysis were:

(1) Apple, which has followed the iPad with the highly successful new generation versions of the iPhone.
(2) The search giant Google, which has moved into the smartphone world with their Android operating system.
(3) Toyota, which is working on an even more fuel-efficient Prius and is developing plug-in gas-and-electric cars.
(4) GE, which has moved into the world of green engineering and redesign of manufacturing technology to make products more suitable for use in low-technology economies.

(5) Microsoft, which is moving into touch screen technology with the launch of Surface and expansion of their smart phone and cloud computing operations.
(6) Tata, the Mumbai-based conglomerate, which has developed the world's cheapest car.
(7) Nintendo, the video game maker, which has used their Wii console to tap an entirely new gaming audience with products such as the Wii fitness game.
(8) P&G, which has outperformed competitors to exploit technology based upon external collaboration to enter new consumer market segments.
(9) Sony, with electronic consumer goods using Blue-ray to enhance the online content of video games such as those available on Playstation 3.
(10) Amazon.com, which is expanding their web services by using outsiders to create online games, entering the downloadable video and music markets and selling Web management services to other companies.

Technology-based innovation usually occurs through scientific research leading to the development of new operating processes, products or services. As illustrated in Figure 9.1, success is influenced by an organisation's knowledge management capabilities, understanding of markets and technological competences. Success involves organisational learning from prior experience, undertaking R&D, staff development, analysis of failures and understanding competitors' successes. Technological competences relate to the ability to utilise infrastructure facilities to support research, testing, concept development and production of the final innovation outcome.

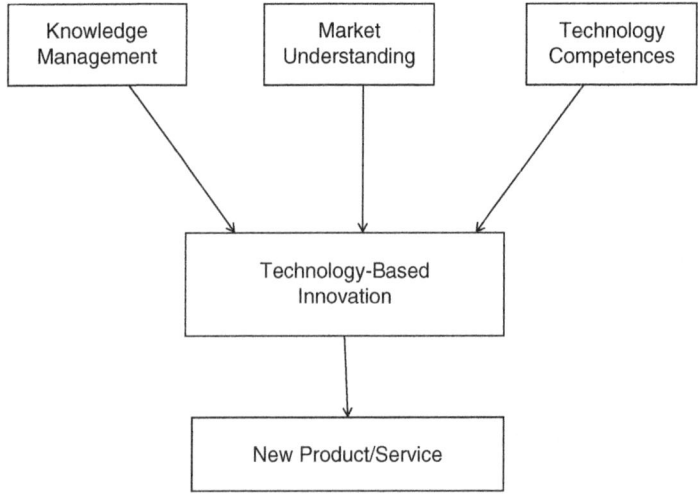

Figure 9.1 Factors of influence

Garg and Garge (2005) suggested technological innovation does have certain similarities with the evolutionary processes found in the natural world. This is because scientific discoveries generate new concepts that provide the basis for modifying old systems and developing new systems. The survival and success of these new systems is then determined by the market accepting the new system or remaining loyal to older concepts. These authors noted however that technological and biological evolutionary processes do have their differences. These include:

(1) Technological innovation is usually motivated by some need or opportunity and the solution is always constrained by some boundary conditions such as customer acceptance or prevailing market conditions. Biological evolution process is caused by random mutations which will be tested by nature's 'survival of the fitness' laws.
(2) The results of technological innovation can be pre-conceptualised in a form before the new process or product is given a final shape. In biological evolution as random mutations, hence final forms often remain unpredicted and un-visualised until they emerge as a physical shape.
(3) Technological innovation is recursive, being predominantly dependent upon designers' experience, accumulated knowledge and working environment in a specific environment. In the case of biological evolution the random nature of mutations means differences will occur but conscious control over outcomes is not possible.

In terms of factors influencing technological innovation Garg and Garge's examination of innovation strategies for IBM, 3M and Hewlett Packard revealed the importance of team work, utilisation of Total Quality Management (TQM), exploitation of leading edge technologies, strong support for innovation by senior management, promotion of individuals based upon the success in previous innovation projects and collaborating with other organisations in the acquisition of new knowledge. The authors noted however that the relative importance given to these various factors does differ between these three organisations.

Benner and Tushman (2003) proposed that breakthrough innovations can be classified on the basis of their (i) advances of existing technology and (ii) departure from the existing conventions within market segments. Zhou et al (2005) defined the advances approach as 'technology-based innovations' and the departure approach as 'market-based innovation'. This latter type of innovation typically involves new and different technologies that meet the requirements of new customers or newly emerging markets. Zhou et al examined the performance implications of firms fostering technology-based innovation versus market-based innovation and the benefits of exhibiting an entrepreneurial orientation. They concluded that in stable or mature markets, marketing orientated firms are able to base their technology innovation upon seeking to satisfy identified existing customers' latent needs. However this approach is likely to fail in new or emerging markets. In these situations firms are advised to accompany their market orientation with adoption of an entrepreneurial focus because this encourages convention challenging and enables the firm to escape the myopia of assuming existing outputs can provide adequate revenue flows in the future.

Responding to market change

Case Aims: To illustrate how technological innovation can be used to switch market emphasis during a downturn in demand from traditional customers.

For several years, mulch producers and composting facilities in North America have faced intense competition for their raw material from companies seeking feed stocks for biomass power plants and wood pellet manufacturing. However when the US construction industry went into recession, demand for lumber fell dramatically, accompanied by a sharp fall in the production of sawdust and bark. This created major problems for the manufacturers of equipment for mulching and composting. However those manufacturers, such as Bandit Industries, who have continued to invest in innovation have been rewarded by increased sales (Goldsten, 2010).

Founded in 1983, Bandit Industries' primary focus was on tree care applications. The emergence of the biomass energy market provided the company with a completely new range of opportunities for supplying equipment to land clearance, demolition and cleaner construction companies. Consequently Bandit developed their Beast models, which are machines that handle whole logs for conversion into biomass energy materials. The machines only produce conventional size wood chips but not the much finer ground materials such as sawdust that are required by the manufacturers of wood pellets. To meet the needs of the pellet industry, Bandit exploited new technology to permit their Beast Recycler machine to produce bio-sawdust from debarked round wood trees. The technology has also been used to offer equipment capable of processing materials such as rice straw and sugar cane used in ethanol production. This latter experience proved to the company that rice straw and sugar cane can also be utilised to produce a fine sawdust that can be offered as a new raw material for pellet manufacturing. Bandit has also developed a metal extraction device which permits removal of contaminants prior to recycling of used wood roofing shingles.

Another area where the economic downturn has been bad news is in the factory automation industry with major manufacturer firms shelving plans for upgrading machinery or opening new facilities (Kirwet,

2010). Festo AG based in Esslingen is a traditional German, family owned company whose long-term view includes continued R&D investment even during difficult trading conditions. Their ability to sustain sales reflects the fact that their customers are seeking ways of using leading-edge technology to improve manufacturing productivity. The other factor permitting ongoing success in the face of price competition is the company's ability to innovate at the total system level by adapting both pneumatic and electric-automation technologies. This has resulted in their machines being more energy efficient, thereby reducing their customers' operating costs.

Importance

Erikson et al (1990) proposed that technology can be divided into three types:

(1) *Base technologies*, which are fundamental to the production of products or services in an industrial sector.
(2) *Key technologies*, which are critical because they provide sources of competitive advantage.
(3) *Pacer technologies*, which can be expected to evolve into the key technologies of the future.

The importance of technology varies by industry in part due to the nature of the product offering. Opportunities to utilise new technology are significantly lower in most consumer food products than, for example, in consumer electronics goods. Hence, as illustrated in Figure 9.2, any analysis of future trends should take into account the degree to which technology permits the firm to (i) achieve a higher level of performance relative to competition and (ii) provide the basis for generating differences in the scale of added value that is achievable within an industrial sector.

Where technology neither permits achievement of superior performance nor greater value, the market is likely to be very fragmented, containing numerous firms all seeking non-technological-based ways of achieving competitive advantage. This scenario can be contrasted with situations where specialist firms can identify an opportunity to exploit technology in order to achieve superiority. In these latter cases the role of technology will be to permit firms to occupy specialised market niches where there are few or no competitors.

Branded goods companies tend to exhibit the ability to achieve greater value through superior marketing competence but may be unable to exploit technology as a strategic pathway for achieving performance superiority. Success is often achieved by an early market entrant capturing the majority of the initial customers and then relying upon competences in areas such as marketing and logistics to sustain their leadership position. This strategy is somewhat different

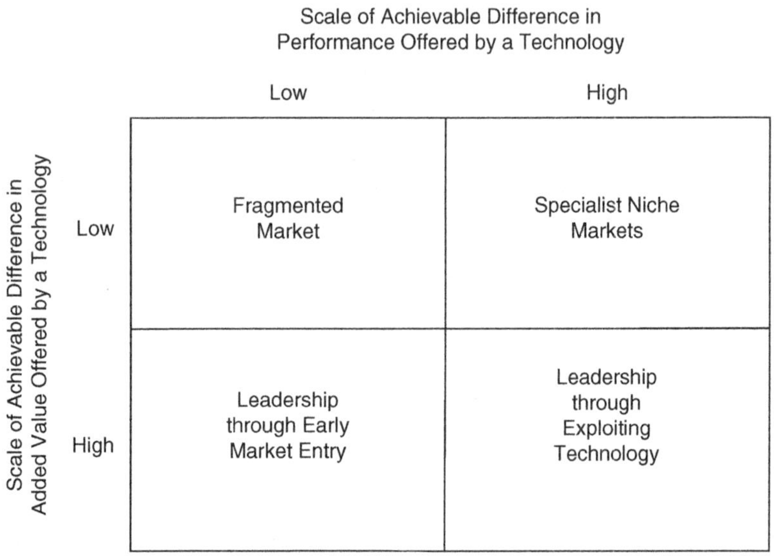

	Scale of Achievable Difference in Performance Offered by a Technology	
Scale of Achievable Difference in Added Value Offered by a Technology	Low	High
Low	Fragmented Market	Specialist Niche Markets
High	Leadership through Early Market Entry	Leadership through Exploiting Technology

Figure 9.2 Alternative market opportunities matrix

where there are major opportunities for the exploitation of technology advances to deliver superior performance or greater product value. Although being an early entry is highly advantageous, in order to retain leadership an organisation will have to continue to invest in new or improved technologies in order to effectively block the attempts of competition to utilise the same new technology as a path through which to match performance.

To be successful scientific discovery and technological innovation must be accompanied by market insight. For example brain scanners were first conceived from a system developed to X-ray metals. Market insight was required to recognise the opportunities for using the technology in a way which revolutionised the market for medical diagnostic equipment. Maximising return on investment (ROI) from R&D usually occurs through focusing upon developing new technologies, rather than concentrating on emerging or mature technologies. Although it may seem appealing to wait until an emerging technology has evolved into a commercially viable opportunity, excessive delay may result in the organisation losing out to firms who have led the emergence of a new technology. Many companies often prefer to stay with the technology they have relied upon in the past. As a consequence there is a tendency to sustain investment in developing existing technology and to avoid what is perceived as the higher risk proposition of expenditure on an emerging technology (Brown, 1992).

Retaining leadership in the innovation stakes

Case Aims: To illustrate that technological innovation demands a long-term commitment to seeking to retain market leadership through superior capability.

A long lead time can exist between concept identification, completion of fundamental research and the ability to launch a new product. An example of being prepared to make this level of commitment in technological innovation is provided by the Japanese car manufacturer Toyota. Long before the American or European car manufacturers exhibited any concerns over rising oil prices, Toyota had the strategic insight to change vehicle transportation from a dependence on hydrocarbons to utilise other types of fuels. Their first product was the highly successful hybrid the Toyota Prius. Since Toyota launched the Prius there has been continuous innovation to improve this vehicle and to expand the company's hybrid product line (Rapp, 2007).

Even in the current period of economic austerity, Toyota has continued to exhibit a long-term perspective in relation to technological innovation. A key aspect of their strategy is to exploit a highly integrated IT system which permits optimal access to the knowledge embedded in their design group, production systems and global automated consumer ordering system. This knowledge has enabled the company to continuously reduce the cost of the hybrid engine and develop fuel cell technology faster than its competitors; thereby ensuring retention of both technological innovation and cost leadership within the car industry.

Toyota's fundamental operating philosophy principle has always been to build 'better products at lesser costs'. To this end, Toyota has developed unique production systems designed to eliminate all forms of waste. The knowledge of all individuals within the organisation is exploited to the fullest through emphasis on improving personal job roles and working environments. As nations seek to respond to global warming, Toyota has focused on exploiting technological innovation to manufacture environmentally friendly cars offering lower emissions and improved fuel economy. More recently the company has exploited Japan's leadership in consumer electronics to evolve the car into a mobile telecommunications device. The ultimate aim is to equip their new cars with a communications platform and a smart function capability to enhance car safety and to optimise energy consumption.

Focus

Retaining market leadership is usually determined by the scale of superior performance delivered by a firm's products. In most cases commitment to innovation will be supported by the organisation's ability to exploit key and pacer technologies. Superior technological competence tends to be a scarce and expensive resource. Consequently when determining how internal resources should be allocated across the organisation, there is a need to identify those markets which represent the greatest source of opportunity. In terms of assessing this issue, one approach is to assume there are two dimensions to be considered; markets and product performance. As illustrated in Figure 9.3, this approach results in an assessment based across four different market/product scenarios.

Schmidt and Porteus (2000) proposed that the winner in a technology-based market battle will be influenced by the relative level of competence of other organisations in terms of cost leadership and technological capability. In the case where two organisations have relatively low levels of both competences, the incumbent will probably adopt a retrenchment strategy whilst the new entrant will invest in new technology. The former's hope is that the new entrant will fail to commercialise the new technology. Where both firms have relatively high levels of competence, the existing firm is likely to retain dominance for existing products while concurrently pursuing opportunities offered by the new technology. This formidable market position will probably cause the competitor to decide there is no benefit in investing in

		Market	
		Existing	New
Product	Existing	Technological Change Influencing Existing Product(s) Performance or Value	Technological Change Influencing Existing Product(s) Adapted for New Markets
	New	Technological Change Leading to a New Generation of Product(s)	Technological Change Leading to Radical Market Change and/or Diversification

Figure 9.3 Product/market change matrix

the new technology and look elsewhere for an attractive investment opportunity. These authors concluded that this latter scenario is exemplified by Intel, the world's leading developer and manufacturer of microchips. Intel has a history of concurrently investing in the development of next generation chips whilst focusing upon driving down the production costs for their current generation of products.

Another issue requiring assessment is the priority given to technological investment to upgrade existing internal organisational processes versus concentrating upon new product development (Fox et al, 1998). Where there is confidence that current process technologies will remain appropriate for the foreseeable future, the company can focus on enhancing current market performance. Should new technology permit the creation of the next generation of products, action will be necessary to invest in product development. Where there is a requirement for concurrent focus on process and product innovation, this more complex situation may demand consideration of a fundamental change in future strategy.

Managing complex alliances

Case Aims: To illustrate some of the issues associated with innovation involving alliances between organisations.

The complexity of modern technology has led more firms to adopt an alliance-based approach to innovation. The pharmaceutical industry is placing increasing reliance upon alliances. This has become necessary because many of the established industry giants lack the necessary capabilities, in the increasingly important emerging area of biotechnology and genomics, to develop new approaches to providing healthcare treatments.

Fuchs and Krauss (2003) posited that small biotech firms which are forming alliances with larger pharmaceutical companies are unique for three reasons. Firstly they are heavily science-based, more nimble and less risk averse than large companies. As a consequence innovation within biotech firms tends to be more radical than in other industries (Powell et al, 1996). Secondly the biotech companies are a major source of tacit knowledge. Thirdly these firms face a very long timeline between establishing the company and the development of a commercially viable product.

Once a new innovation has proved viable, small biotech companies usually lack the resources to manufacture and market their new product and will need to identify other organisations with whom to partner to bring their technological innovation to market. Although capable at exploiting new technology, many lack the skills to effectively manage relationships with large partners. These weaknesses are the prime reason why many small biotech firms eventually fail (Gassman et al, 2004).

Shaista et al (2006) proposed that traditional, stage gate linear new product development models are not relevant when managing complex technological innovation. This is because the process is constituted of complex processes involving basic research, innovation and invention, early-stage technology development, product development, and production and marketing. Furthermore within the process, in the case of the biotech industry, there are various critical stages in the development cycle such as granting or patents, government agency approvals and clinical trials. The outcome of each of these steps will determine whether ongoing financing can be obtained from venture capitalists or other partners within an alliance. During the pre-discovery and invention stages the lack of evidence of commercial viability often means the biotech firm is reliant upon government grants. Investors are only interested once a tangible commercial product has been identified. Final validation of commercial opportunity can only occur after a lengthy approval process involving various government agencies.

Shaista et al concluded that the following paradoxes will confront the small company:

Paradox 1: Is the market willing to pay for the costs associated with commercialisation of the idea?
Paradox 2: Can successful alliances be formed without running the risk of trade secrets being lost?
Paradox 3: Can the alliance be maintained when problems emerge that indicate that potential success is unlikely unless complex problems can be resolved?
Paradox 4: Can the project move ahead with the innovation even though the biotech firm lacks the managerial skills to supervise commercialisation?
Paradox 5: How much control should be granted to commercial partners and how much control should be retained by the biotech firm?
Paradox 6: How much ownership should the biotech firm be prepared to lose in order to gain access to investment funds?

Building effective alliances is a significant challenge for biotech firms. The available evidence suggests small biotech companies should identify common goals and use formalised contracts to define clear-cut roles and expectations for the alliance. Shaista et al concluded that attempts to utilise a linear sequential development model will cause the biotech

> firm to adopt inflexible organisational mechanisms and exhibit a lack of understanding of the complex forces which exist within projects involving collaboration with larger organisations. Success can only be achieved when the biotech firm understands the factors influencing relationships, the interactivity of the science and technology required to develop a commercially viable product and the organisational capabilities to successfully develop and launch a new healthcare treatment.

Technology transfer

Successful large scale technological innovation such as the development of a new aircraft relies on the innovation capabilities of upstream suppliers. This is because the final product is reliant upon a host of subassemblies and components. Adner and Kapoor (2010) proposed that this situation not only demands an ability merely to manage innovation inside the lead organisation, but requires competences among all partners to handle the challenges which exist in this type of project. Such scenarios should be treated as 'ecosystems' in which equal focus is given to managing bottlenecks and comprehending the magnitude and specific innovation outcomes that are demanded.

Most partnership-based major technological innovation is undertaken under the direction of a 'focal firm'. The challenges faced by this firm depend upon whether problems are created by upstream or downstream participants. The former scenario is likely to create many more problems for the focal firm than the latter. Adner and Kapoor suggested technological challenges tend to decrease as technologies mature, but that contractual challenges tend not to weaken over time. They posited that the existence of bottlenecks is evidence that challenges are distributed unevenly across the firms engaged in the project. Upstream component problems will constrain value creation because the focal firm's ability to produce the final product can be impaired. This can be contrasted with downstream bottlenecks, because these constrain the customer's ability to derive full benefit from the focal firm's product.

Where the focal firm's final new product is very new or complex, Adner and Kapoor suggested that final end users may be dependent upon the activities of downstream 'complementors'. An example is the new Airbus A380 Super Jumbo, where airlines' adoption is dependent upon major airports creating the ground facilities required to handle such a large aircraft. In those cases where complementors move rather slowly, these delays will increase the probability that other companies will be able to develop and launch a competitive offering; thereby threatening the potential benefits of being a first-mover organisation.

An increasing proportion of successful firms' revenues is derived from technology discovered outside the firm (Scherer, 2010). The success of inbound technology transfer is determined by a firm's 'absorptive capacity', which is an ability to recognise, assimilate and exploit in the context of innovation and organisational learning. Lichtenthaler and Lichtenthaler (2010) proposed that another determinant of success is 'prior knowledge', because this enables a company to identify valuable external knowledge in order to effectively utilise this input to enhance innovation. Opportunity also exists for firms to transfer technology outwards to other organisations. Transfer success is influenced by 'desorptive' capacity which Lichtenthaler and Lichtenthaler defined as an organisation's ability to identify outbound technology transfer opportunities and to facilitate the technology's application by the recipient. The authors noted that outward technology transfer does not preclude internal technology application. This is because firms may license some of the technological knowledge also concurrently being utilised within their own product portfolio.

Royalty generation will be influenced by the breadth of market knowledge of the knowledge owner. Thus firms which passively await licensing opportunities will generate less revenue than more active organisations. For example, a business unit at Dow Chemical developed a new glue and patented an automotive application. The unit's narrow market knowledge was subsequently shown to have precluded the opportunities for licensing the technology to numerous areas outside the car industry (Davis, 2008). A high desorptive capacity can enhance the firm's inward technology transfer success because technology transactions often involve mutual technology transfer in both directions through activities such as cross-licensing agreements or R&D alliances (Davis and Harrison, 2001).

Teece (1986) concluded that in a world of increasing technological complexity an innovator's profits are now often predicated by access to complementary assets held by others. Nevertheless there is evidence that even where collaboration is a preferred strategy, some firms still remain concerned that this approach risks proprietary or confidential knowledge leaking out of their partner organisations and becoming known to competitors.

Increasingly firms involved in technological innovation are relying upon their membership of business networks as a strategy to access new knowledge. Ernst (2005) concluded that participation in networks has a critical role in assisting firms to exploit innovation as the basis for moving from a cost-leadership to a differentiation strategy. This is because the network can provide access to new knowledge about market conditions and the opportunities to use innovation to exploit emerging market trends.

Liu and Wu (2011) posited that firms which minimise the number of redundant or non-productive ties with other network members are more able to access opportunities to acquire new knowledge. These authors believe this outcome will be reflected in superior performance when implementing a differentiation strategy based upon achieving superior product performances through innovation. Their perspective is supportive of Uzzi's (1997) view that network membership will increase the speed of new knowledge acquisition, enhance the quality of decision making and lead to improved performance.

Stable industry technology opportunities

Case Aims: To illustrate that even in highly stable industries opportunities exist to exploit technology to make products more affordable.

Food products represent one of the largest areas of consumer spending and during an age of austerity there will be increasing pressure to exploit technology to increase the affordability of products. Two primary areas of opportunity are to reduce the energy costs associated with manufacturing processes and to extend shelf life, thereby reducing prices and in-store product wastage levels. Fryer and Versteeg (2008) reviewed the new technology currently under development within the food industry to identify future opportunities to reduce manufacturing costs or extend shelf life.

Microwave and radio frequency energy can be used to rapidly heat foods. Although in use in the home and the catering sector, adoption within the processing sector industry is still very limited. This is because of the need to overcome the tendency of microwaves to unevenly heat the product, which severely limits food safety and food spoilage. Ohmic heating involves passing electric current through the food which then becomes hotter as a result of food's inherent electrical resistance. As with microwave heating, the technical challenge is to obtain thermal uniformity. The process is more energy-efficient than microwaves but requires good electrical contact with electrodes whilst avoiding product contamination.

High pressure processing can lengthen the fresh qualities of food by subjecting materials to intense pressures which de-activates microbes and viruses. High pressure thermal sterilisation combines pressure and thermal processes to sterilise food. Both approaches are likely to come on stream over the next few years to extend shelf-life and remove the need for chemical preservatives. Pulsed electric field (PEF) is a non-thermal process that uses very short, high voltage pulses. The technique can de-activate microbes in liquid foods without reducing the flavour and colour. Applications of PEF under development include doubling of shelf life of pasteurised milk, producing microbially safe, cold pasteurised fresh juices and reducing the costs and time taken to dry materials prior to their use as ingredients in processed products.

Ultrasonics processing, which involves the use of sound waves, has been utilised for cleaning surfaces and containers for years. Projects now in progress include reducing the energy costs associated with the mixing of ingredients and enhancing fermentation processes to increase speed and expand plant capacity. Cool plasma technology involves the use of a gas such as helium, oxygen or nitrogen with molecules being 'excited' by the application of an electrical field. When the gas particles come into contact with micro-organisms, they lose their charge and this deactivates the micro-organisms. The benefit of cool plasma technology is that it can be used at low temperatures.

S-curve

Not all customers or organisations respond at the same speed in terms of adopting a new technology. This behaviour in relation to the percentage of customers adopting a new technology may result in an S-shaped product adoption curve of the type shown in Figure 9.4 (Ortt and Schoorman, 2004). The shape of the curve is influenced by three different phases within the adoption process. Firstly there is the 'development phase' during which the new technology is identified, developed and evolved into a feasible

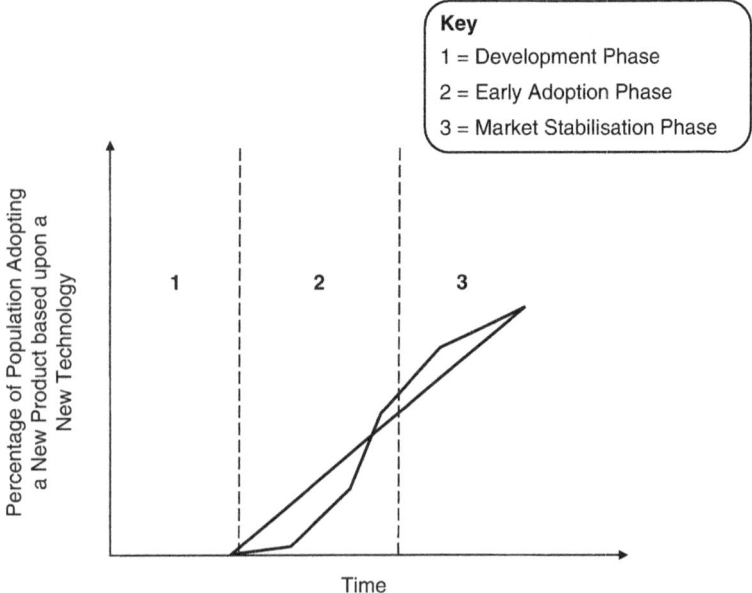

Figure 9.4 The innovation S-curve

proposition of appeal to potential users. This is followed by the 'early adoption phase' during which the more innovative, risk-taking members of the adopting population are prepared to incorporate the new technology into their ongoing purchasing activities. Once the new technology has clearly demonstrated a benefit by the early adopters, the more conservative elements within the population will commence purchasing because they now perceive there are few risks in also becoming users. Ultimately all of the later market entrants will have adopted the new technology and the total number of users will plateau.

In assessing the scale of potential opportunities associated with the expected shape of the S-curve for a new technology, a critical issue is the speed with which the new customers adopt the new technology (Brown, 1992). One of the obstacles confronting the introduction of a new technology is the cost of the new product or service relative to the price of goods available from firms using existing technologies. This is because new technology is often expensive to develop and it may only be feasible to launch the first generation goods by charging a relatively high price. As organisations gain experience with a new technology, this usually leads to a decline in production costs, which then permits a price reduction. As prices fall, new customers may enter the market. This will enable new firms to enter the market by exploiting their capabilities in areas such as managing high-volume production systems or their experience in the marketing of mass-market goods. Where a reduction in the cost of a technology is not feasible, product prices will remain high and the probable outcome that is the number of customers will remain low.

Benkenstein and Bloch (1993) proposed that the nature of the technology within an industrial sector will influence the nature and the shape of S-curves. In markets characterised by stable technologies, the S-curve tends to be elongated, relatively shallow and reflective of long life cycles. The intensity of innovation competition is relatively low and technology management is of less importance that other operational issues. Where there is substantial technological change within a sector but no technological discontinuities are occurring, the S-curve tends to be steep and profitability from investing in R&D is above average. There is constant pressure to introduce new state-of-the-art products. As a consequence innovation is a critical success factor and life cycles tend to be relatively short. This places pressure on firms to constantly seek new ways of sustaining competitiveness.

In turbulent technology situations, technological discontinuities occur frequently and old technologies are overtaken by new advances. Sudden leaps in technology will obsolete knowledge and production assets. The priority task is to accurately identify new technologies which offer the largest gains in product or process performance improvement or, conversely, which represent a potential technological bottleneck. It can also be the case that success is dependent upon the simultaneous development of several technologies. This latter situation demands very high R&D management capabilities (von Braun et al, 1990).

A concept associated with the marketing of new products is the 'diffusion of innovation', which proposes potential customers can be divided into five groups; namely 'innovators', 'early adopters', 'the early majority', 'the late majority' and 'laggards'. The time taken for a potential customer to first purchase a product will depend upon which group the individual is a member of. The first purchases will be by innovators, whereas the last individuals to ever purchase will be laggards. Having examined the launch of high technology products, Moore (1991) concluded that the benefits sought by of each of the five customer groups are somewhat different. Innovators purchase the product because they wish to own the latest technology, being prepared to accept any problems which may exist with the new product. Early adopters will need to be persuaded that the product will work properly and can offer a new way of fulfilling their vision of wishing to exploit a new technology. The early and late majority will postpone purchase until they are persuaded the product offers a functional benefit not provided by existing products. Laggards are price sensitive and hence wait until the product is virtually obsolete before entering the market.

Moore used case materials from a number of product launches to demonstrate that these different needs will require a change of product benefit offered to customers as companies seek to 'cross the chasm' which exists between the five customer types. Moore's chasm theory suggests that unless the product benefit is revised to reflect different market needs, then at each phase along the diffusion of the innovation curve there is the risk that the new product will not attract the next group of customers. This scenario represents both an opportunity and a threat. The threat is that a currently successful firm fails to develop an effective new benefit proposition and is unable to cross the next chasm. This outcome is an opportunity for another firm, should this latter organisation be more able to deliver the benefit sought by the next customer group.

The inability to cross the next chasm due to inappropriate competences can be illustrated by examining competitive developments in the electronic calculator market. The early leader was Hewlett-Packard (HP) selling complex and expensive products to its traditional customers. HP lacked the capabilities to develop and market a lower cost proposition capable of appealing to non-scientific users. Texas Instruments with competences in both the manufacture of large volume goods and mass marketing skills was successfully selling to this segment of the market and the wider general public. However Texas Instruments lacked the competences to manufacture and market a very low priced product capable of exploiting the lower end of the consumer mass market. As a consequence the Japanese manufacturer Casio, which had this expertise, was able to enter the market and displace Texas from its leadership position (Brown, 1992).

References

Adner, R. and Kapoor, R. (2010), Value creation in innovation ecosystems, *Strategic Management Journal*, Vol. 31, pp. 306–333.

Benkenstein, M. and Bloch, B. (1993), Models of technological evolution: their impact on technology management, *Marketing Intelligence & Planning*, Vol. 11, No. 1, pp. 20–28.

Benner, M.J. and Tushman, D. (2003), Exploitation, exploration, and process management: the productivity dilemma revisited, *Academy of Management Review*, Vol. 28, No. 2, pp. 238–56.

Brown, R. (1992), Managing the 'S' curves of innovation, *The Journal of Consumer Marketing*, Vol. 9, No. 1, pp. 61–73.

Davis, L. (2008), Licensing strategies of the new 'intellectual property' vendors, *California Management Review*, Vol. 50, No. 2, pp. 6–30.

Davis, J.L. and Harrison, S.S. (2001), *Edison in the Boardroom: How Leading Companies Realize Value from Their Intellectual Assets*, John Wiley & Sons, New York.

Ernst, D. (2005), Pathways to innovation in Asia's leading electronics-exporting countries—a framework for exploring drivers and policy implications, *International Journal of Technology Management*, Vol. 29, No. 1/2, pp. 6–20.

Fox, J., Gann, R., Shur, A., Von Glahn, L. and Zaas, B. (1998), Process uncertainty: a new dimension for new product development, *Engineering Management Journal*, Vol. 10, No. 3, pp. 19–27.

Fryer, P.J. and Versteeg, C. (2008), Processing technology innovation in the food industry, *Innovation: Management, Policy & Practice*, Vol. 10, No. 1, pp. 74–90.

Fuchs, G. and Krauss, G. (2003), Biotechnology in comparative perspective, in: *Biotechnology in Comparative Perspective*, Fuchs, G. (ed.), Routledge, New York, pp. 1–13.

Garg, S. and Garge, S.B.L. (2005), Technology innovation as an evolutionary process, *Global Journal of Flexible Systems Management*, Vol. 6, No. 1, pp. 41–50.

Gassman, O., Reepmeyer, G. and Zedwitz, M.V. (2004), *Leading Pharmaceutical Innovation: Trends and Drivers for Growth in the Pharmaceutical Industry*, Springer-Verlag, New York.

Goldsten, N. (2010), Competition for feedstocks: wood recycling and processing trends, *BioCycle*, Seattle, January, pp. 38–40.

Kirwet, A. (2010), Downturn spurs innovation, *Machine Design*, January 14th, pp. 29–30.

Lichtenthaler, U. and Lichtenthaler, E. (2010), Technology transfer across organizational boundaries, *California Management Review*, Vol. 53, No. 1, pp. 154–165.

Liu, X. and Wu, X. (2011), Technology embeddedness, innovation differentiation strategies and firm performance: evidence from Chinese manufacturing firms, *Innovation: Management, Policy & Practice*, Vol. 13, pp. 20–35.

McGregor, J. (2008), Most innovative companies: smart ideas for tough times, *Business Week*, New York, April 28th, pp. 62–64.

Moore, G.A. (1991), *Crossing the Chasm*, The Free Press, New York.

Ortt, J.R. and Schoorman, J.P.I. (2004), The patterns of development and diffusion of breakthrough communication technology, *European Journal of Innovation Management*, Vol. 7, No. 4, pp. 292–230.

Powell, W., Koput, K. and Smith-Doerr, L. (1996), Interorganization collaboration and the locus of innovation: networks of learning in biotechnology, *Administrative Science Quarterly*, Vol. 41, pp. 116–145.

Rapp, W.V. (2007), Hydrocarbons to hydrogen Toyota's long-term IT-based smart product strategy, *The Business Review*, Cambridge, Vol. 7, No. 2, pp. 1–7.

Scherer, F.M. (2010), Pharmaceutical innovation, in B. Hall and N. Rosenberg (eds), *Handbook of the Economics of Innovation*, North Holland, Amsterdam, pp. 539–574.

Shaista E.K., Mroczkowski, T. and Bernstein, B. (2006), From invention to innovation: toward developing an integrated innovation model for biotech firms, *Journal of Product Innovation Management*, Vol. 23, pp. 528–540.

Schmidt, G.M. and Porteus, E.L. (2000), Sustaining technology leadership can require both cost competence and innovative Competence, *Manufacturing & Service Operations Management*, Vol. 2, No. 1, pp. 1–19.

Teece, D.J. (1986), Profiting from technological innovation, *Research Policy*, Vol. 15, No. 6, pp. 285–306.

Uzzi, B. (1997), Social structure and competition in interfirm networks: The paradox of embeddedness, *Administrative Science Quarterly*, Vol. 42, No. 1, pp. 35–67.

Von Braun, C.F., Fischer, H.G. and Muller, A.E. (1990), The need for and the issues involved in integrated R&D planning in large corporations, *International Journal of Technology Management*, Vol. 5, pp. 559–576.

Zhou, K.N., Yim, C.K. and Tse, D.K. (2005), The effects of strategic orientations on technology- and market-based breakthrough innovations, *Journal of Marketing*, Vol. 69, pp. 42–60.

10
Strategy Implementation

Influencers

A number of factors influence the success and failure of innovation strategies. Available information permits generic conclusions to be reached concerning the most critical issues. Key factors of influence include whether the innovation is (i) incremental or radical, (ii) aimed at consumer versus B2B markets and (iii) based upon exploitation of complex technology. Van der Panne et al (2009) concluded the most important factors include:

(1) The firm's culture being strongly committed towards innovation and management perception that future performance is dependent upon exploiting innovation.
(2) The firm has extensive experience in managing innovation.
(3) Innovation is driven by a multidisciplinary team with appropriate balance between technological and marketing skills.
(4) Each innovation project has the support of senior management.
(5) A clearly defined strategy exists concerning the focus of innovation.
(6) The innovation is compatible with the organisation's core competences.
(7) The innovation creates a relative advantage of performance, quality or price.
(8) The marketing launch effectively executed and internal departmental conflicts avoided.
(9) Excessively rigid decision making or hierarchical management structures are avoided.

Many new ideas originate from an in-depth understanding of markets and customers. Some researchers believe excessively close interaction between organisations and markets can reduce entrepreneurial creativity by causing the firms to focus on imitating competition or only developing improved versions of existing products. Hence firms must decide whether their future is to be determined by continuous versus radical innovation, plus selecting the most appropriate operating philosophy and organisational structure. There is only limited agreement among researchers about the importance of top management support. In some studies innovators sustained their activities without the blessing of the senior management team. Differences of opinion also exist over whether firms which invest heavily in R&D enjoy a greater level of success. These differences may reflect that in some situations rapid rate of customer acceptance within a market is more critical than technological leadership.

Effectiveness of teams and departments is essential to strategy implementation. Where strategies remain unchanged, accumulated experience and skills will minimise potential errors; whereas a strategy reliant upon radical innovation may encounter problems because employees are unfamiliar with some of their new roles. Klein and Sorra (1996) proposed that assessing potential strategic capability should involve determining whether organisational culture ensures employees are committed and supportive towards the proposed strategy. Importantly when employees' concerns and complaints arise regarding the planned innovation, management must respond in order that any potential obstacles to successful innovation are removed before the strategy implementation phase commences.

Klein and Sorra concluded that employees who perceive innovation is congruent with their work values are more likely to be committed and enthusiastic. Employees' values reflect the degree to which people share common aims and purpose. These shared values evolve as a result of shared work experiences and personal characteristics (Schein, 1992). Group values reflect the self-interests of teams, departments and different levels within the managerial hierarchy. Differences are usually a function of a group's (i) role in the organisation, (ii) common interactions and experiences and (iii) knowledge and skills (Dougherty, 1992). A poor fit between individual and group values may result in entrepreneurial activities being ineffective when responding to the environmental change (Hattrup and Kozlowski, 1993). García-Morales et al (2006) posited that an inward-looking organisation will be unable to proactively respond to environmental change. Proactivity permits the organisation to implement the innovation required to ensure ongoing success.

Katz and Alien (2004) concluded that organisational structure will influence strategy implementation in relation to the degree to which an organisation is mechanistic (exhibiting lower complexity, higher formalisation and centralisation) or organic (possessing ways to organise for creativity and innovation). Organic structure tends to be more appropriate for facilitating actions such as delegated responsibility, work force empowerment and lower level staff having access to required resources without firstly gaining approval from senior managers.

Dewett et al (2007) noted that a more functionally differentiated firm is created where different technologies are utilised in different units of an organisation. This characteristic benefits the speed with which technological innovations will be adopted. These authors also believe professionalism is important because professionally trained employees are more receptive to adopting new technology. Nevertheless division of labour in large organisations can result in excessive formalisation and less contact between technical and administrative personnel, leading to obstacles to progress unless management ensure effective internal communication systems exist.

High sector exit costs may present a barrier to innovation implementation. When an existing technology is accepted as a standard within a sector the lead innovator risks being the sole adopter of a new technology leading to sectoral isolation within a market (Afuah, 1998). Even where exit costs are lower, a firm can expect to incur significant costs in switching technology. Katz and Alien (2004) proposed that to ensure resources and staff exist to sustain current operations and concurrently manage the switch to a new technology, management must establish 'slack resources' such that resource allocation decisions can be optimised.

McAdam (2005) noted that innovation implementation involves the three key constructs of normative evaluation, legitimisation and conflict. Normative evaluation is concerned with assessing proposed innovation within an organisation's conventions, routines and practices. Legitimisation involves a sense-making process of integrating or rejecting the proposed innovation at a group or organisational level in the company. Conflict can occur where the innovation is perceived as breaking widely accepted industry conventions. Massey (2001) proposed the following recursive stages within the process of conflict management and legitimisation:

(1) Destabilise legitimisation of existing conventions.
(2) Attract stakeholder support for alternatives.
(3) Gain organisational acceptance for alternatives.
(4) Gain legitimacy for alternatives such that these evolve into the organisation's new conventions.

Managing innovation

Case Aims: To illustrate how different organisations approach the management of implementing an innovation strategy.

Larsson and Bergfors (2009) proposed that management of innovation strategies will be influenced by whether the focus is on product or process innovation designed to reduce production costs or improve productivity. The authors examined innovation in three European firms: Arla Foods, Billerud and Boliden.

Arla Foods is Europe's largest dairy company with a turnover in the region of $8 billion and employing 20,000 people. The company is a co-operative owned by approximately 13,500 milk producers in Denmark and Sweden. At Arla Foods about 85 percent of the R&D budget is concerned with product innovation. The company is divided into three autonomous divisions; Nordic, International and Ingredients, each with their own research centre. The Nordic centre focuses upon product innovation concerning fresh milk and desserts, the International centre works on cheese and butter and the Ingredients division concentrate their interests around milk powder. Over 200 new products are launched each year with the marketing focus being that of striving to beat competition. Knowledge created in a research centre is distributed to the other research centres through workshops and formal meetings. In the area of process innovation the large production technology suppliers to the dairy industry have been unwilling to come up with process innovations which satisfy the needs of Arla Foods. For this reason, Arla Foods has now decided to establish an internal group to work on process innovation. The aim is to develop knowledge to reduce production costs across all of the company's processing plants.

Billerud is a packaging paper company with sales of $960 million that employs 2,600 people in 11 countries. At Billerud about 50 percent of the R&D budget is focused on product innovation and 50 percent on process innovation. Within Billerud product innovation initiatives were previously introduced by technicians at each paper mill. Following concerns that the company's new products lacked market-orientation, senior management redefined product innovation focus as being Market Pulp, Packaging Boards and Packaging & Speciality Paper. Process innovation is conducted in isolation at each paper mill, focusing upon issues relevant to that specific plant. There is no formalised cross-plant co-ordination and systems for knowledge sharing about process innovation.

Boliden is a mining and smelting company producing copper, zinc, lead, gold and silver. Annual sales are in the region of $3 billion and the organisation employs approximately 4,500 individuals. Boliden does not engage in product innovation. The smelters extract and produce basic metals from ores and the challenge for the company is to produce standard products using raw materials of varying quality. Each production plant is autonomous, which means there is no long-term focus on how innovation might be exploited to sustain overall organisational performance.

Larsson and Bergfors concluded that product and process innovation may be organised on a centralised or decentralised basis. At Arla Foods both product and process innovation is centralised. This reflects a desire to ensure new knowledge is communicated across the company to guarantee delivery of radical innovation outcomes. At Billerud product innovation is centralised while process innovation is decentralised. Production plant autonomy reduces any incentive to pursue process innovation on a company-wide basis. A similar situation in relation to process innovation exists at Boliden.

Radical innovation

In an age of new austerity more firms will probably need to rely on radical innovation. This will permit companies to restructure markets, redefine supply chain economics and create entirely new product categories (Abdul, 1994). Large, established firms are very capable of managing operational efficiencies and improving products or services. They tend to encounter difficulties in managing radical innovation because this activity often involves technical or market uncertainties. Technical uncertainties reflect problems over the validity of the underlying science or whether the technology will prove commercially viable. Market uncertainties include issues over customer needs and willingness to switch to very different product or service propositions.

By tracking 12 radical innovation projects, Leifer and Rice (2001) identified two other sources of uncertainty; namely resources and organisational capability. Resource problems may arise because the organisation has insufficient resources or, alternatively, management has not ensured existing operations are making available resources needed to support radical innovation projects. Organisational uncertainties include issues such as the project team capability, relationships across the organisation, overcoming a short-term, results-orientated orientation of and counteracting any vested interests in retaining the organisation's current business model.

The researchers' recommendation for overcoming these constraints is to create a 'radical innovation hub'. This serves as mentoring source during idea generation, concept development, advising project teams over resource acquisition and managing interfaces with existing business units and senior management. Leifer and Rice also identified the following additional factors that can enhance the success rate for radical innovation:

(1) Appointing individuals to identify new ideas from within and outside the organisation. These individuals must have the capability to think about scientific breakthroughs and advances in technology connections, and understand the nature of social trends, markets and customer needs.

(2) Appointing project leaders who are able to overcome uncertainty obstacles and develop viable solutions.
(3) Ensuring access to resources to avoid project teams spending an inordinate amount of time overcoming resource constraints.
(4) Effective project transition management whereby the radical innovation can be developed into a commercially viable solution that provides the basis of a new business unit.
(5) Appointing individuals prepared to be risk takers, exhibit drive and think out-of-the-box.
(6) Forming multifunctional teams to ensure an adequate breadth of thinking at all stages in the innovation process.
(7) Identifying visionary champions among senior management in order that radical projects are supported at the top level within the organisation.

Chandy and Tellis (2000) undertook a review of new product development programmes in American firms to identify sources of radical innovation. They determined that 42 percent of radical innovation came from large incumbent firms, which appears to contradict the conventional view that non-incumbent firms are a more probable source of innovation. In relation to firm size, the study found that 58 percent of radical innovation was by small firms. This result is supportive of the idea that small firms are more able to exhibit more entrepreneurial behaviour. Nevertheless given that large firms are the source of 42 percent of identified radical innovation, it would appear this latter group are also an important source of different products or internal production processes. The data also revealed that large firms account for a substantially larger proportion of radical innovations relative to their total number in the US economy than do small firms.

Chandy and Tellis (1998) concluded that a high proportion of dominant firms in today's high-technology industries are willing to cannibalise past investments in order to introduce radical product innovations ahead of the competition. Furthermore within US high-technology operations there has been a move towards decentralisation, to enhance flexibility and speed of response. Many companies now operate as a group of autonomous organisational units, which enables retention of an entrepreneurial orientation but concurrently permits these units to benefit from access to resources from across the organisation. Another benefit enjoyed by large firms is diverse technological capability, which ensures early recognition of scientific breakthroughs and an ability to rapidly resolve any technology problems should these arise. Chandy and Tellis suggested that small firms do need to comprehend the reality that limited resources and access to diversity of technological knowledge can frustrate radical innovation. They suggested there are at least two possible solutions. One is to exploit research spill-over from more resource-rich firms or public sector research centres and the other is to partner with other organisations to gain access to required technological capabilities and resources.

Outsourcing innovation

Case Aims: To illustrate the issues associated with having a third party undertake innovation on behalf of the organisation.

As technology has become more complex firms are finding that certain aspects of innovation need to be outsourced to entities such as research institutes, universities, suppliers and small technology start-up companies. Cui et al (2009) reviewed case materials on the activities of the large German company Siemens. This company has a huge in-house R&D capability employing over 32,000 staff. Nevertheless the company is increasingly using outsourcing where access to external knowledge of new-to-the-world technology is perceived as beneficial.

A common factor in successful projects is the high degree of commitment and trust which exists between Siemens and their outsourcing partners. Whether the project can be outsourced or not is determined in part by the potential partner having the necessary competences. In the case of universities and research institutes, these partners must have a reasonable understanding of the commercial realities associated with developing a financially viable solution. With supplier partners, the emphasis is on ensuring they can develop technically optimal solutions in the shortest possible time. Thus suppliers must have strong in-house R&D competence to overcome technical problems that may arise. When working in partnership with customers the key issue for Siemens is evidence of the customer's extensive and relevant market knowledge.

Knowledge complexity

Chiva et al (2007) proposed the key dimensions influencing innovation in a complex knowledge situation are (i) participative decision making, (ii) experimentation, (iii) knowledge management, (iv) risk taking and (v) internal dialogue. Mat and Razak (2011) proposed that participative decision making is important because there is a need to ensure all the relevant parties are contributing to the process. Additionally participation in decision making can reduce resistance to change within the organisation (Senge, 1990). Thomke (1998) concluded that experimentation maximises the flow of ideas which challenge established conventions and thereby ensure adequate organisational creativity. Experimentation will be found to be a core value within companies which are successful at implementing radical innovation strategies. This is because in many cases it is only through the use of trial and error that solutions to complex problems will emerge.

Successful innovation involving complex technologies may involve interaction with other organisations within the market system. Thus employees must have the competences to exchange knowledge with others from outside the organisation (Alavi and Leidner, 2001). Accompanying exploitation of external knowledge, successful innovation will involve a willingness to take risks. Successful innovative companies appear more willing to engage in risk taking when compared to their less innovative counterparts (Covin et al, 2006).

Dialogue is an important contributor in building inter-functional collaboration, cohesiveness, trust and commitment among different functional areas across the organisation (Balthasar et al, 2000). Achieving effective inter-functional coordination is critical because complex knowledge will usually demand the integration of specialist capabilities from across the organisation. Within open innovation, organisational learning should extend to those outside the company who are contributing to the innovation process. Lichtenthaler (2011) made a distinction between different forms of knowledge. He proposed that 'external knowledge exploration' is the activity concerned with acquiring new knowledge from external partners. Storage of knowledge is defined as 'internal knowledge retention' and can be contrasted with 'external knowledge retention' which is knowledge maintained by external partners.

Lichtenthaler and Lichtenthaler (2009) posited that firms need to develop 'inventive capacity' which involves the ability to utilise internal knowledge, generate new knowledge and integrate this new knowledge into existing knowledge bases within the organisation. 'Absorptive capacity' refers to the ability to explore and understand external sources of new knowledge (Zahra and George, 2002). The ability to translate new external knowledge into understanding of benefit to the organisation's innovation programmes is known as 'transformative capacity' (Lane et al, 2006). Existence of 'connective capacity' ensures ongoing access to privileged or proprietary information. Determining which internal knowledge can be made available to others is known as 'desorptive capacity'. This must be effectively managed so that others can gain maximum mutual benefit whilst the provider avoids leakage of competitive advantage within the market system.

For open innovation to occur there is a requirement for 'complementarity' between the partners (Chesbrough et al, 2006). The interfaces and interactions associated with developing effective complementary relationships will be strongly influenced by the attitudes of those involved in the collaboration. Gavetti (2005) noted that inappropriate attitudes or poorly developed interpersonal skills may inhibit successful relationships, thereby creating barriers to effective project implementation. Implementation goals and competences must be aligned with corporate strategy. In those cases where the ability to engage in radical innovation requires external links, these must be established by working with partners who have a proven entrepreneurial track record.

> ### Extended innovation
>
> *Case Aims: To illustrate various ways whereby collaboration can assist in the implementation of an innovation strategy.*
>
> Muller and Valikangas (2002) proposed that declining transaction costs, due to factors such as the utilisation of IT to permit inter-firm communication, have reduced the economy of scale benefits which companies derive from internally fulfilling all business processes such as innovation, procurement, manufacturing, sales and distribution. As organisational boundaries have become more porous firms are finding that collaborative alliances provide the basis for 'extended innovation'.
>
> Extended innovation appeals to firms in high-technology industries because of the difficulty and expense for a single firm to own all of the necessary competences to develop new-to-the-world products. For this reason extended innovation has become popular in the healthcare industry. University laboratories engage in pure research and then collaborate with biotechnology firms which undertake the applied research to convert a scientific discovery into a viable technology. This knowledge is then shared with the large pharmaceutical companies who undertake the large-scale production and distribution required to market the new drug.
>
> Some firms in mature industries have begun to realise that extended innovation involving firms from other industry sectors may be beneficial. One example is provided by the US retailer Home Depot. The company is collaborating with real estate developers in Southern California to publicise the latter's residential developments. In return the developers are permitting Home Depot to demonstrate painting techniques, window treatments and landscape designs in their show homes. Home Depot and insurance company Allstate Insurance are collaborating in a process to direct householders to the retailer's stores. Allstate adjusters encourage contractors to buy replacement materials from Home Depot stores and this assists Allstate to reduce the size of claims for home repair costs.

Collaboration

Sivadas and Dwyer (2000) concluded that a large number of alliances fail due to the following factors:

(1) Lack of trust due to members seeking to protect their own proprietary knowledge whilst attempting to gain access to proprietary knowledge of other alliance members.
(2) Lack of effective communication between partners.

(3) Lack of effective co-ordination such that outcomes are not achieved within previously agreed time schedules.
(4) Lack of ability to generate the new knowledge required to support the proposed innovation.
(5) Lack of new product management development competences within one or more of the members of the alliance.
(6) Lack of reciprocal interdependence due to inadequate understanding of what each partner should undertake or inadequate core competences to fulfil assigned areas of responsibility.

Creemer (2008) noted that effective knowledge exchange is critically dependent upon task co-ordination and the sharing of learned experiences. To achieve this goal there is benefit in creating procedural control agreements. Creemer concluded that when the degree of dependency increases partners will strive to achieve positive outcomes because partners are less inclined to use power or to behave opportunistically. High levels of dependence also lead to mutual relationships becoming more important, with partners more willing to respect each others' opinion and to accept compromises in order to reach an agreement over key issues. Where each of the partners has a specialist role and utilise proprietary knowledge, it is more difficult for partners to validate each others' claims over progress being achieved. Under these circumstances, Creemer determined, the success of alliances is heavily reliant upon all parties exhibiting a high level of integrity and mutual trust.

During the 1990s as firms sought to enhance productivity and optimise operational costs, in many market systems there was a move towards greater co-operation between different members of industry supply chains. Simatupang and Sridharan (2002) proposed supply chain collaboration strategies can be of a horizontal, vertical and lateral nature. Horizontal integration involves two or more organisations at the same level of the supply chain producing similar products or different components of one product sharing their resources such as manufacturing capacity. Benefits can include (i) reduced logistics and administration costs, (ii) improved procurement terms through greater buying power, (iii) lowering of the fixed costs and (iv) improved access to markets by being able to offer greater continuity of supply. Vertical integration involves collaboration between participants at different levels within the supply chain. Lateral collaboration combines the benefits and sharing capabilities of both vertical and horizontal integration.

Tether (2002) suggested the strategy which offers greatest opportunity to firms within the same supply chain is 'innovation co-operation'. This involves participation in joint R&D and other technological innovation activities. The benefits include accessing resources not available within a single organisation, risk reduction and more effective marketing. Walters and Rainbird (2007) concluded that providing complementary knowledge and user know-how permitted (i) improvements in the balance between price and product performance, (ii) drawing upon user expertise to refine the benefit proposition and (iii) achieving greater market awareness.

Creation of more efficient and effective supply chains has resulted in collaboration being directed towards process innovation whereby together organisation can create, produce, communicate and deliver greater value to the final customer. This activity has been greatly assisted by the advent of IT and the Internet which have permitted organisations to exchange data in real time, analyse demand patterns to optimise product flows and make services available at the time demanded by customers. Malhotra et al (2005) concluded that information sharing and building of information technology infrastructures also allows participants to create new knowledge and enables development of new products or services. The US computer firm Dell has developed strong, information rich relationships with suppliers facilitating the creation of the competitive advantage whereby customers can specify unique product specifications at time of purchase and then track in real time the progress of their order passing through the supply chain.

Walters and Rainbird (2007) proposed that opportunities available from collaborative innovation can be assessed using value chain analysis. The approach permits identification of added-value activities both inside and outside an organisation. Johns et al (2005) noted however that this achievement requires the ability to overcome the problems associated with co-ordination, communication and overall control across the entire supply chain. There must also be genuine commitment by all participants to focus on innovation that benefits all members of the whole supply chain instead of seeking opportunities that merely benefit the individual firm. To achieve this outcome, Walters and Rainbird propose that all members of the supply chain must understand:

(1) What drives industry competitiveness.
(2) What the customer value drivers are and what the scope is to add additional value.
(3) How added-value opportunities can be used to achieve mutual competitive advantage.
(4) How co-operative innovation can be used to create added value.
(5) What the roles and responsibilities of each of the supply chain partners are.
(6) What resources are required to achieve a positive outcome from co-operative innovation.
(7) What data demands must be met to manage partner interactions.

Absorptive capacity

Case Aims: To illustrate the need for compatibility in the absorptive capacity of those engaged in open innovation.

Lance (2010) posited that, in supply chain open innovation, each downstream partner must perceive value in working with upstream members. For this to occur, each partner requires the absorptive capacity to comprehend how other contributors are providing an opportunity for value generation. Absorptive capacity is a learning capability constituted of the four processes of knowledge acquisition, assimilation, transformation and exploitation (Zahra and George, 2002).

Lance examined how absorptive capacity influences final outcomes by examining the development of a new drug treatment. The initial scientific breakthrough occurred within a university laboratory. To convert this new knowledge into a potentially viable treatment, the scientist needed to involve a company. The effectiveness of this relationship was influenced by a number of factors, including:

(1) The downstream customer's ability to utilise the new scientific knowledge to develop a value proposition.
(2) This firm's deployment of absorptive capacity processes in contributing specialised expertise.
(3) This firm's abilities to collaborate with the upstream source of the new scientific knowledge.

The downstream company lacked the capabilities to engage in the manufacture of the new drug or to market the new drug to the healthcare sector. Hence there was a need to involve another downstream partner in the project. At this juncture the downstream company faced a role reversal: moving from customer for an idea to being a supplier of a proposition of interest to another partner capable of marketing the new drug. This outcome required the necessary absorptive capacity to attract a customer interested in progressing development of the new drug. Concurrently this final customer's absorptive capacity needed to be compatible with the supplier's absorptive capacity, because different questions, mindsets, relationships and goals will influence attitudes and behaviour of the various partners in a supply chain relationship. Success is also dependent upon whether the downstream customer responsible for the launch of the new products has the capability to identify the potential end users and can present a proposition whereby these potential purchasers perceive added value from adopting the new product. This will only occur where there is compatibility between the absorptive capacity of the supplier of the new treatment and those in the healthcare sector engaged in the treatment of relevant illnesses.

In the case analysed by Lance the commercial team within the supplier organisation failed to recognise the value offered by the new drug. This situation adversely affected the intensity of effort they exerted in developing and marketing the new drug. As a consequence insufficient energy was expended on seeking compatibility with the absorptive capacity of end users, such as doctors and national health schemes. This failure ultimately resulted in much lower than expected sales.

References

Abdul, A. (1994), Pioneering versus incremental innovation: review and research propositions, *Journal of Product Innovation Management*, Vol. 11, pp. 56–61.

Afuah, A. (1998), *Innovation Management Strategies, Implementation, and Profits*, Oxford University Press, New York.

Alavi, M. and Leidner, D.E. (2001), Review: knowledge management and knowledge management systems: Conceptual foundations and research issues, *MIS Quarterly*, Vol. 25, No. 1, pp. 107–136.

Balthasar, A., Battig, C. and Wilhelm, B. (2000), Developers-key actors of the innovation process, *Technovation*, Vol. 14, No. 2, pp. 523–269.

Chandy, R.K. and Tellis, G. (1998), Organizing for radical product innovation: the overlooked role of willingness to cannibalize, *Journal of Marketing Research*, Vol. 34, pp. 474–487.

Chandy, R.K. and Tellis, G. (2000), The incumbent's curse? Incumbency, size, and radical product innovation, *Journal of Marketing*, Vol. 64, No. 3, pp. 1–17.

Chesbrough, H., Vanhaverbeke, W. and West, J. (2006), *Open Innovation: Researching a New Paradigm*, Oxford University Press, Oxford.

Chiva, R., Alegre, J. and Lapiedra, R. (2007), Measuring organisational learning capability among the workforce, *International Journal of Manpower*, Vol. 28, No. 3/4, pp. 224–242.

Covin, J.G., Green, K.M. and Slevin, D.P. (2006), Strategic process effects on the entrepreneurial orientation-sales growth relationship, *Entrepreneurship Theory and Practice*, January, pp. 57–81.

Creemers, P.A. (2008), Alliance governance and product innovation project decision making, *European Journal of Innovation Management*, Vol. 11, No. 4, pp. 472–487.

Cui, Z., Loch, C.H., Grossmann, B. and He, R. (2009), Outsourcing innovation, *Research Technology Management*, November/December, pp. 54–72.

Dewett, T., Whittier, N.C. and Williams, S.C. (2007), Internal diffusion: the conceptualizing innovation implementation, *Competitiveness Review*, Vol. 17, No. 1/2, pp. 8–25.

DiPadova, L.N., Faerman, S.R. and Human, S. (1993), Using the competing values framework to facilitate managerial understanding across levels of organisational hierarchy, *Human Resource Management*, Vol. 32, No. 1, pp 143–174.

García-Morales, V.J., Llorens-Montes, F.J. and Verdú-Jover, A.J. (2006), Antecedents and consequences of organizational innovation and organizational learning in

entrepreneurship, *Industrial Management + Data Systems*, Vol. 106, No. 1/2, pp. 21–42.
Gavetti, G. (2005), Cognition and hierarchy: rethinking the microfoundations of capabilities, *Organization Science*, Vol. 16, No. 6, pp. 599–617.
Hattrup, K. and Kozlowski, S.W.J. (1993), An across-organization analysis of the implementation of advanced manufacturing technologies, *Journal of High Technology Management Research*, Vol. 4, pp. 175–196.
Johns, R., Crute, V. and Craves, A. (2005), Improving value delivery: challenges in establishing value chain delivery, *Proceedings 2nd European Forum on Market-Driven Supply Chains*, Milan, April, pp. 56–67.
Katz, R. and Alien, T.J. (2004), Organizational issues in the introduction of new technologies, in Staw, B.M. (ed.), *Psychological Dimensions of Organizational Behavior*, Vol. 3, Prentice-Hall, Upper Saddle River, NJ, pp. 450–463.
Klein, K.J. and Sorra, J.S. (1996), The challenge of innovation implementation, *The Academy of Management Review*, Vol. 21, No. 4, pp. 1055–1081.
Lance, N. (2010), Wearing different hats: how absorptive capacity differs in open innovation, *International Journal of Innovation Management*, Vol. 14, No. 4, pp. 703–731.
Lane, P.J., Koka, B.R. and Pathak, S. (2006), The reification of absorptive capacity: a critical review and rejuvenation of the construct, *Academy of Management Review*, Vol. 31, No. 4, pp. 833–863.
Larsson, A. and Bergfors, M.E. (2009), Product and process innovation in process industry: A new perspective on development, *Journal of Strategy and Management*, Vol. 2, No. 3, pp. 261–276.
Lichtenthaler, U. and Lichtenthaler, E. (2009), A capability based framework for open innovation: complementing absorptive capacity, *Journal of Management Studies*, Vol. 46, No. 8, pp. 1315–1338.
Lichtenthaler, U. (2011), Open innovation: past research, current debates, and future directions, *Academy of Management Perspectives*, Vol. 25, No. 1, pp. 75–93.
Leifer, R. and Rice, M. (2001), Implementing radical innovation in mature firms: the role of hubs, *The Academy of Management Perspectives*, Vol. 15, No. 3, pp. 102–113.
Malhotra, A., Gosain, S. and El Sawy, O.A. (2005), Absorptive capacity configurations in supply chains: gearing for partner-enabled market knowledge creation, *MIS Quarterly*, Vol. 29, No. 1, pp. 145–187.
Massey, J. (2001), Managing organisational legitimacy: communication strategies for organisations in crisis, *The Journal of Business Communication*, Vol. 38, No. 2, pp. 153–182.
Mat, M. and Razak, R.C. (2011), The Influence of organizational learning capability on success of technological innovation, *International Journal of Business and Social Science*, Vol. 2, No. 7, pp. 217–226.
McAdam, R. (2005), A multi-level theory of innovation implementation: Normative evaluation, legitimisation and conflict, *European Journal of Innovation Management*, Vol. 8, No. 3, pp. 373–388.
Muller, A. and Valikangas, L. (2002), Extending the boundary of corporate innovation, *Strategy & Leadership*, Vol. 30, No. 3, pp. 4–9.
Schein, E.H. (1992) *Organizational Culture and Leadership*, Jossey-Bass, San Francisco.
Senge, P. (1990), The leaders new work: building learning organizations, *Sloan Management Review*, Fall, pp. 7–23.
Simatupang, T.M. and Sridharan, R. (2002), The collaborative supply chain, *International Journal of Logistics Management*, Vol. 13, No. 1, pp. 15–30.

Sivadas, E. and Dwyer, F.R. (2000), An examination of organizational factors influencing new product success in internal and alliance-based processes, *Journal of Marketing*, Vol. 64, No. 1, pp. 31–49.

Tether, B. (2002), Who co-operates for innovation, and why: an empirical analysis, *Research Policy*, Vol. 31, pp. 13–22.

Thomke, S. (1998), Managing experimentation in the design of new products, *Journal of Management Science*, Vol. 44, pp. 743–762.

Van der Panne, G., Van Beers, C. and Kleinknecht, A. (2009), Success and failure of innovation: a literature review, *International Journal of Innovation Management*, Vol. 7, No. 3, pp. 309–338.

Walters, D. and Rainbird, M. (2007), Cooperative innovation: A value chain approach, *Journal of Enterprise Information Management*, Vol. 20, No. 5, pp. 595–607.

Zahra, S.A. and George, G. (2002), Absorptive capacity: a review, reconceptualization, and extension, *Academy of Management Review*, Vol. 27, No. 2, pp. 185–203.

11
The Service Sector

Managing services

Confronting any service organisation seeking to satisfy customers is the problem that the goods supplied are often intangible. Mechanisms available that provide some form of tangibility include:

(1) *Place*, which is the physical setting around which the provision of services is delivered.
(2) *People*, who are involved in working with the customer–organisation interface.
(3) *Equipment*, of the necessary standard to rapidly and efficiently assist in supporting the service provision process.
(4) *Communication systems*, to inform stakeholders of the organisation's role in the provision of services.

Another characteristic of services is inseparability. This describes the fact that many services are simultaneously produced and consumed. Hence for many service outcomes to occur, both the provider and the customer must be able to interact with each other and this requires adequate staff resources to support customer interactions. This objective can be difficult when an organisation faces declining market demand and there is a need to reduce employee headcount. Where this occurs the organisation should seek to find alternative service delivery mechanisms. One solution is to create online operations offering features such as downloadable forms, smart system guidance software to assist form completion, an online call centre and a website providing online purchasing facilities.

Service demand can be variable in nature because of diverse customer needs and differing service delivery capabilities of employees within the provider organisation. To manage variability, front line staff need training in (i) efficiently handling simple enquiries and (ii) ensuring more complex service requests are passed on to an appropriately qualified person within the organisation.

Unlike manufactured goods, services are highly perishable. Sasser (1976) proposed a number of strategies for responding to this problem, including:

(1) *Differential pricing*, where the customer is charged a lower price for access to services at less busy times.
(2) *Alternative service provision*, to meet the needs of customers during peak periods.
(3) *Service modification*, to ensure that during peak periods the needs of major customers receive priority.
(4) *Demand management systems*, permitting services provider to rapidly (i) identify current available capacity and (ii) implement alternative solutions.
(5) *Temporary capacity expansion*, involving an increase in capability to respond to customer demand during peak periods.
(6) *Service sharing*, where a number of organisations work together and are willing to cross-refer clients.
(7) *Customer participation*, whereby customers are encouraged to become self providers.

Targeted promotions

Case Aims: To illustrate how service firms are exploiting new technology to enhance sales promotion effectiveness.

During any economic downturn service firms will probably turn to sales promotions to attract new users and retain customer loyalty. A major drawback in the past was having only limited knowledge of customer behaviour, which resulted in poorly targeted promotional campaigns. Electronic data acquisition at point of purchase and the Internet have assisted firms to become more sophisticated in their use of sales promotions, thereby permitting activities to be more cost effective. Access to real-time promotional effectiveness data allows service firms to analyse customer and competitor behaviour and, where necessary, introduce revised, more effective promotions (Neslin and Shankar, 2009).

Targeting activities now utilise the use of data from loyalty programmes to identify customer purchasing history as the basis for delivering customised promotions. Technology enablers that are being used to achieve this outcome include mobile phone apps, online personal shopping assistants and in-store kiosks. In the case of online personal shopping, the employees have access to shoppers' purchase history and can generate personal shopping lists consisting of accurately targeted offers.

In the USA the drugstore chain CVS offers targeted promotions based on its Extra Care loyalty programme data. The retailer is able to categorise customers into several segments, target promotions for each segment and disseminate information about promotions through personalised e-mail and other communications. Promotional effectiveness is assessed by using matched control groups to assess impact upon customer purchasing behaviour. Where loyalty programme data is unavailable, one solution is to work with companies such as Catalina who can develop targeted coupon promotions using other data sources on customer shopping habits (Gambardella et al, 2011).

Mobile Internet usage is growing and, as more consumers use smartphones, this has resulted in an increase in mobile marketing by service firms (Shankar et al, 2011). This type of marketing offers two-way communication, causing service firms to add online promotions to their integrated marketing communications by using short message services (SMS).

US retailers such as Gilt, RueLaLa and HauteLook now offer a restricted range of fashion products to select customer groups who must subscribe to the firms' websites. Sometimes subscriptions can only exist when the customer is invited by another subscriber. This can lead to a sense of exclusiveness more highly valued by consumers than inclusive promotions (Barone and Roy, 2010). The added benefit of 'invitation-only' promotions is a reduced chance that a large number of consumers will see the offer and form adverse opinions or expect exclusive fashion items to be available at discounted prices. Another approach is online 'conditional' promotions where a specified condition has to be met for the consumer to have access to the promotion. Lee and Ariely (2006) researched the situation where promotional uptake is under the control of the consumer (e.g. a discount only becomes available when total purchases exceed a specified value). They concluded that these promotions are very effective when consumers have less concrete shopping goals. Other conditional promotions are based upon a condition beyond the control of the consumer. For example, a free coffee offered by Dunkin Donuts when a popular sports team is playing or refunds made available by a retailer should a certain team win a major championship (Bortman, 2009).

The advent of mobile marketing has led to utilisation of group buying websites by service firms to engage in volume-based promotional activities that involve offering a larger discount when a predetermined number of consumers agree to purchase goods. Two of the most

> popular websites are Groupon.com and LivingSocial.com. Consumers can register online to receive relevant deals and coupons. They can decide whether to take advantage of each deal and must do so within a specified time period. To ensure the minimum number of other users sign up for the deal, users rely on their own e-mail contacts and social media to encourage others in their personal networks to participate (Boehret, 2010).

Service quality

The unique properties associated with service goods of intangibility, heterogeneity, inseparability and perishability caused academics to conclude that, in many markets, achieving differentiation relies upon offering superior service quality (Teas, 1994). Understanding customer needs and the organisation's ability to offer a unique service proposition requires that service quality can be measured (Parasuraman et al, 1985). Service quality, Grönroos (1984, 1993) proposed, consists of the following three components:

(1) *Functional quality*, constituted of the process related attributes of behaviour, attitude, accessibility, appearance, customer contact, internal relationship and service-mindedness.
(2) *Technical quality*, constituted of the output related attributes of employees' technical ability, employees' knowledge, technical solutions and computerised systems.
(3) *Image* of the service provider encompassing the customer's general perceptions of the service provider's attributes and delivered benefits.

Parasuraman et al (1985, 1988, 1991) focused upon the concept that service quality can be assessed by examining whether a gap exists between customers' expectations and perceptions. These researchers posited that customers base their assessment of service quality on the five dimensions of reliability (an ability to perform the promised service dependably and accurately), responsiveness (a willingness to help customers and provide prompt service), assurance (employees' knowledge and an ability to inspire trust and confidence), empathy (evidence of caring, individualised attention) and tangibility (the nature of the physical attributes of the provider). They utilised these five dimensions to develop their SERVQUAL scale for measuring service gaps.

The key aim of the service organisation is to ensure perceptions of service are equal to or exceed actual experience. The services gap model permits identification of potential quality problems which can arise during the management of service provision (Zeithmal and Bitner, 1996). The nature of potential gaps includes:

Gap 1: where organisations have not acquired accurate information about what customers expect.
Gap 2: where organisations fail to monitor the service delivery process.
Gap 3: where organisations lack the internal capabilities to deliver the required standard of service.
Gap 4: where incorrect information is communicated to customers.
Gap 5: which represents overall combined impact of the problems created by Gaps 1 through 4.

Lack of adequate customer purchasing data can cause service firms to assume customers are seeking to purchase standardised services. Electronic data from sources such as credit cards, store checkouts and online buying provide today's service firms with extensive knowledge of different customer segments. Organisations can now offer customised services reflective of diversified service needs. An ongoing failure to appreciate the increasingly diversified service needs of different customers in a period of austerity does mean that utilisation of the basic services gap model may not be effective. This can arise because the firm has not recognised that (i) heterogeneous customer groups exist and (ii) customers' expectations can change during a long economic downturn. Failure to recognise either of these two trends may result in an inability to fulfil customers' service quality expectations. The potential consequences, as summarised in Figure 11.1, are the emergence of the following service gaps:

Gap 1a: where the organisation has not acquired accurate information about different customer groups' expectations.
Gap 1b: where the organisation does not have up-to-date information about changing customer expectations.
Gap 2a: where the organisation fails to set quality standards reflecting expectation variation.
Gap 2b: where the organisation is slow to update quality standards to reflect the current quality expectation of different groups.
Gap 3a: where the organisation fails to develop the internal capabilities required to deliver differentiated standards of service.
Gap 3b: where the organisation is slow up-skilling internal capabilities required to deliver ongoing changing expectations.
Gap 4a: where standard information is communicated to customer groups who have different information requirements.

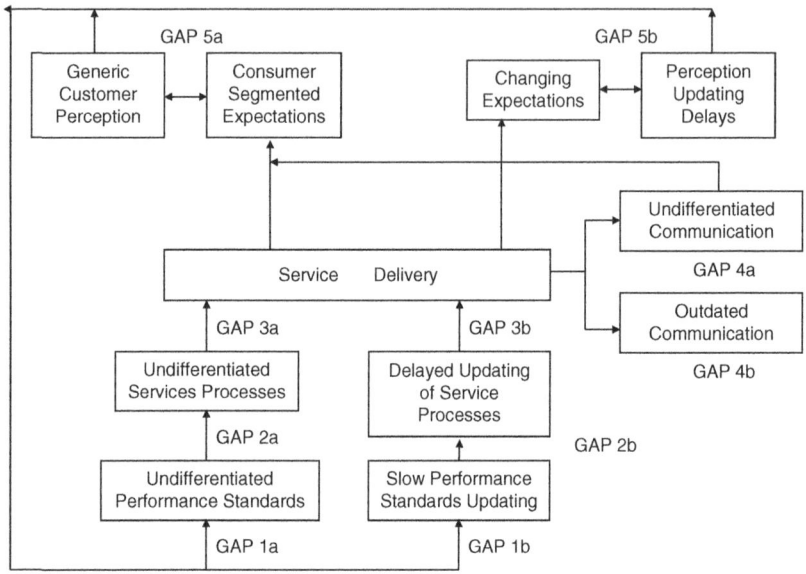

Figure 11.1 Diversified consumer market service gap model

Gap 4b: where outdated information is being communicated.
Gap 5a: representing the combined impact of the problems created by Gaps 1a through 4a.
Gap 5b: representing the combined impact of the problems created by Gaps 1b through 4b.

> **Service quality in austerity**
>
> *Case Aims: To illustrate the issue that the importance of service quality may decline during a period of austerity.*
>
> The US sub-prime mortgage problem, the UK government taking over certain banks and the European sovereign debt crisis have combined to create a liquidity crisis in the banking industry. Banks have needed to reconsider their lending priorities. In the UK this has resulted in small firms facing difficulties obtaining new loans from their banks (Dominic, 2011) and government intervention to persuade the banks to make more financial support available. One outcome was Project Merlin whereby the banks agreed to meet specific small business lending targets (Brownsell, 2011; Kaletsky, 2011).

Despite the existence of Project Merlin there is evidence that the UK banks are still not keen on making loans available to small firms (Anon, 2011). One survey by a trade association found that 60 percent of owner/managers still do not trust their banks following the financial crisis and the subsequent tightening in lending policies. Only 28 percent of businesses reported that they have been able to secure funding from their banks and as a consequence many have been forced to turn to friends or family to secure money to support their operations (Banscombe, 2011). Chief's (2011) research supported the view that, during the current economic downturn, the UK's banks were completely unconcerned about being perceived as delivering inadequate levels of service. This outcome is unlikely to be attributable to managerial incompetence. Hence the question arises whether during an economic downturn conventional thinking about the importance of service quality to sustain customer satisfaction and loyalty are actually perceived as important within the banking sector, despite the fact that these variables continue to be stressed as a critical influencer of performance within the academic literature.

One alternative explanation is that service quality is not perceived as being a key issue within the UK bank sectors. This is possibly reflective of banks needing to reduce balance sheet liabilities created by past inappropriate lending decisions. Another possibility is that within the academic literature too great an emphasis has been placed on the importance of service quality in relation to influencing customer behaviour. Winstanley (1997) concluded that although customers may be dissatisfied with the service experience provided by their bank, in reality customers attach much greater importance to factors such as pricing, convenience and outlet location. King's perspective (1999) was that competition between banks has led to 'commodification' of financial services causing price, not quality of service, to become the dominant determinant of customers selecting their bank.

Sullivan (2008) noted that firms in many consumer service sectors need to accept that the West has moved into a period where 'austerity marketing' has become the prevailing philosophy. This is because greater emphasis is being placed by consumers on the issue of lowest available prices when reaching a purchase decision or deciding to remain loyal to a specific service provider (Grossberg, 2009). Other indicators of the growing importance of price are provided by the rapid increase in consumers utilising 'price apps' on their smartphones when shopping (Lovinson, 2010) and the need for retailers to offer massive price reductions in order to sustain sales during the Christmas period.

Service innovation

The increasing levels of competition among service providers means that innovation is necessary to protect market share (Agarwal et al, 2003). Den Hertog et al (2010, p. 491) defined service innovation as 'a new service experience or service solution that consists of one or several of the following dimensions: new service concept, new customer interaction, new value system/business partners, new revenue model, new organizational or technological service delivery system'. They proposed that opportunities through which to achieve innovation include new service concepts, customer interactions, value system/business partners, revenue models or service delivery systems. In their view, new service concepts can be created through collaboration with the customer to generate new ideas for solving service quality problems or fulfilling unmet customer needs.

Developing a new form of market interaction involves revising the interface between the provider and the customer. Advances in electronic banking such as the introduction of ATMs, online banking and the use of mobile phones are examples where although service levels may be improved, the primary beneficiary has been the banks in reducing their costs of delivering banking services. Innovation involving a new value system relies upon coproduction. In the smartphone sector this approach was used by Apple to market the iPhone. The company used a combination of opening their own iStores and forming partnerships with other software firms to create a range of new apps for the product.

The emergence of new revenue models are somewhat rarer events in service sector industries because in many cases there is the requirement to involve multiple actors in the service provision. One company which has been very successful in achieving this outcome is Amazon; by forming links with other service firms who can become affiliates, marketing their online offerings via the Amazon website. Creating a new delivery system is also not a simple form of innovation as there is often a requirement to persuade customers to alter their purchase behaviour. A retail firm which has accomplished this is the Swedish furniture operation IKEA. At the outset the company had to motivate customers to assemble their own furniture and create an internal culture in relation to how staff should interact with customers visiting IKEA stores. Most forms of technology-based innovations in the service sector have been reliant upon improvements in computer hardware and software. In recent years the Internet has become the primary source of technology-based innovation. In the travel and tourism industry, for example, the Internet has enabled the creation of service customisation and the introduction of self service concepts such as online booking.

Unlike tangible goods innovation, new service concepts can rarely be researched, developed, prototyped and tested before market launch. This is because the conceptual nature of many service innovations makes it difficult for customers to assess what will be experienced or delivered until actual market introduction (Parasuraman et al, 1985). Furthermore where customers are seeking standardised products and minimal price variation between service providers, companies are usually forced to use the same internal operational processes and technologies. Consequently developing new products which fulfil the attributes of being valuable, rare and not easily duplicated by competition can prove to be an impossible task.

Innovative service provision

Case Aims: To illustrate how exploitation of new technology can enhance the effectiveness of the service provision process.

Domino's Pizza is one of the world's leading pizza delivery companies operating in more than 50 countries with approximately 8,000 self-owned and franchised outlets. In the UK, the Domino's brand and business is operated by Domino's Pizza Group Limited. In the pizza market the core proposition includes offering a variety of flavours, developing innovative textures, shapes and crusts and making available a number of beverage or food add-ons. A critical reason for Domino's success is the investment in ongoing service innovation. One of the most successful innovations was the development of the firm's Heatwave box which is designed to make sure Domino's pizzas are delivered hot every time. As well as ensuring the pizza remains hot during the transit, the box also prevents sweating which can ruin the quality of the pizza crust (Anon, 2009).

The company is very aware that success demands a superior ability to ensure customers can order a pizza quickly and with minimum fuss. Most ordering is done over the telephone. This is a potential weak link in the service provision chain due to factors such as a bad telephone line or a customer not having access to an up-to-date menu and price list. To sustain leadership in this aspect of the service provision process, Domino's has invested in creating an online ordering system.

Domino's UK has also launched a nation-wide SMS service whereby customers can place an order using their mobile phone. A textual ordering channel is available for those who have difficulty in ordering over the phone due to reasons such as a lack of product familiarity or limited confidence in using English. Textual order confirmation also helps to avoid ordering mistakes by customers.

Since 2005 customers ordering from a landline can choose their local store or call a national number to be redirected to their nearest store. This same service has now been introduced for mobile phone users, with the customer being directed to the last store they ordered from, or alternatively the Domino's ordering system can triangulate their present location and direct them to the nearest store. By analysing the data from online and mobile phone ordering the company can engage in relationship marketing using promotional techniques such as offering a time-banded voucher which is based upon the customer's proximity to a Domino's store at a time of day which is related to their past purchasing behaviour.

Business model innovation

An organisation's business model determines the value proposition, sources of revenue, the resources utilised and the nature of the links which exist with the organisation's stakeholders (Zott and Amit, 2010). Issues covered by the model should include structure, operational activities, processes that connect the organisation's internal functional activities and the external constituencies, which together provide the basis for delivering a firm's strategy (Teece, 2010). Service markets can cause major problems in identifying and implementing a unique strategy that cannot rapidly be duplicated by competition. As a consequence business models have an important role in supporting superior performance of service market organisations (Sorescu et al, 2011).

Operational efficiency is concerned with undertaking organisational processes in a cost effective, rapid way that permits effective exploitation of resources. One way of achieving operational efficiency is to streamline back-end operations through actions such as improved inventory management or order processing. In the case of service businesses efficiencies may be available through upgrading the front end environment in a way that can reduce costs or increase revenue. Retailers, for example, can generate higher revenues by developing innovative store layouts, merchandise displays and shelf allocation (Murray et al, 2010).

Business model upgrades

Case Aims: To illustrate the benefits of changing business models in service sector operations.

Cost savings in service organisations can be realised by adopting new technologies that automate processes previously handled by employees. For example, the DVD rental supplier Netflix, modified bar-code sorting machines to handle the odd-shaped envelopes used to mail out DVDs, thus increasing the number of envelopes being processed (Stross, 2010).

The American retailer Zappos automated their fulfilment centre by utilising an autonomous robot system resulting in only 12 minutes passing from the time an order is received online to the picking, packing and shipping of an order (Scanlon, 2009).

Zara is, in revenue terms, the world's largest clothing retailer (Bjork, 2010). Initially the firm became successful using standard ageing inventory and brand management techniques. In a world of increasing price competition and declining margins the company recognised the need to evolve a new business model in order to create a competitive advantage. The premise of the new business model is to use a smaller assortment of clothes accompanied by faster-turning stock to create an aura of exclusivity and cut down on the need for regularly announced sales to clear excess inventory. In an industry where merchandising is seasonal, Zara challenged industry conventions by stocking stores with new designs twice a week (Rohwedder and Johnson, 2008). This shake up of retailing practice was enabled by the development of sophisticated operations research models which determine the most efficient manner to distribute inventory from Zara's two central warehouses to over 1,500 stores worldwide (Caro et al, 2010).

Redbox is a US firm which operates a chain of automated kiosks that dispense DVD rentals for $1 per day. The primary focus of the firm's business model is to exploit self-service as the basis for achieving higher operational efficiency. Unlike many other consumer service businesses, Redbox has sought to remove employees from the service delivery process. This has been achieved through the use of new technology to create automated kiosks placed in convenient locations such as McDonald's restaurants and major supermarkets. This high efficiency model permits Redbox to offer an assortment of DVDs at a price significantly lower than those available from the competition (Krauss, 2009).

Effectiveness and efficiency

Service firms often have a distinct advantage over their counterparts in manufacturing industries in that their business model is not defined by a specific production technology. This means the service firm has more flexibility in determining their product assortment and can more rapidly respond to changes in market demand. It may also be possible to exploit demand by 'leveraging complementarities'. An example of a company that has achieved this in the retail sector is Apple, by opening company-owned stores (Useem, 2007). These outlets are unique environments where customers can experience the products, get one-on-one tutorials on a wide range of technical issues, have devices repaired at the Genius Bar or participate in workshops.

Service firms have traditionally sought to increase customer efficiency by offering the product in multiple locations, self-service convenience or staff providing customer support. The advent of the Internet has increased the efficiency of the service provision experience by reducing customers' search costs and allowing the purchase of products that were previously geographically inaccessible. In the case of retailers, the Internet has also enabled selling online, in terrestrial outlets and across different channels by allowing customers to purchase online and pick the product up at a store. Alternatively customers can access the retailer's larger online assortment whilst shopping in store where they can take advantage of customer support (Sorescu et al, 2011).

Business model change aimed at enhancing the customer experience involves the service provider seeking new ways of facilitating customer purchasing activities. Traditionally this was achieved by increasing the breadth of product assortment but this can be extremely expensive in terms of managing inventory or causing conflicting demands over display space. With service providers seeking to sustain profitability during the new austerity, the tendency is to focus on generating revenue from the most popular goods. This may result in some areas of demand remaining unmet or certain smaller customer segments having to be ignored. The Internet offers a solution through making less popular items available by warehousing these in a centralised location. The Internet has also permitted the expansion of product assortments through 'customer co-creation'. This approach to service provision is being adopted across a number of industries. For example, customers can now create their own granola using their favourite ingredients by visiting www.mixmygranola.com, personalise their chocolate M&Ms with text and pictures at awww.mymms.com or design their own sports shoes online using the NikeID system. By exploiting mass customisation this strengthens brand image and increases loyalty by leveraging the psychological ownership effect that arises from customers co-designing products (Franke et al, 2010).

References

Agarwal, S., Erramilli, M.K., and Dev, C.S. (2003). Market orientation and performance in service firms: Role of innovation, *The Journal of Services Marketing*, Vol. 17, No. 1, pp. 68–80.

Anon (2009), Building from a platform of scale and innovation to grow during recession, Datamonitor, London, March, pp. 1–11.

Anon (2011), Smaller UK firms struggle for bank finance, *Euroweek*, London, October 21st, pp. 3–4.

Banscombe, P. (2011), Scottish entrepreneurs don't trust banks, *The Scotsman*, March 15th, p. 3.

Barone, M. and Roy, T. (2010), Does exclusivity always payoff? Exclusive price promotions and consumer response, *Journal of Marketing*, Vol. 74, No. 2, pp. 121–32.
Boehret, K. (2010), A deal on a haircut? That's what friends are for, *The Wall Street Journal*, New York, March 24th, p. D3.
Bjork, C. (2010), Inditex profit jumps On Zara Chain's sales, *The Wall Street Journal*, June 14th, p. B6.
Bortman, Eli, C. (2009), The Jordan's Furniture monster Deal: a legal gamble? *Sport Marketing Quarterly*, December, pp. 218–221.
Brownsell, A. (2011), Banking on better, *Marketing*, London, June 22nd, pp. 17–18.
Caro, F., Gallien, J., Díaz, M., García, J., Corredoira, J.M., Montes, M., Ramos, J.A. and Chief, N. (2011), Illusory Merlin under fire, *The Scotsman*, November 20th, p. 1.
Den Hertog, P., Van der Aa, M. and De Jong, M.W. (2010), Capabilities for managing service innovation: towards a conceptual framework, *Journal of Service Management*, Vol. 21, No. 4, p. 490–514.
Dominic, J. (2011), UK banks fall short of lending targets for businesses, *The Scotsman*, May, p. 35.
Franke, N., Schreier, M. and Kaiser, U. (2010), The 'I designed it myself' effect in mass customization, *Management Science*, Vol. 56, No. 1, pp. 125–140.
Grönroos, C. (1984), A service quality model and its marketing implications, *European Journal of Marketing*, Vol. 18, No. 4, pp. 36–44.
Grönroos, C. (1993). Toward a third phase in service quality research: challenges and future direction, in Swartz, T.A., Bowen, S.W. and Brown, S.W. (eds), *Advances in Service Marketing and Management*, 2nd edn., JAI Press, Greenwich, CT, pp. 33–42.
Grossberg, K.A. (2011), Indulgent parsimony: an enduring marketing approach, *Strategy & Leadership*, Vol. 39, No. 2, pp. 36–42.
Kaletsky, A. (2011), Opinion: how can banks, businesses and government be encouraged to reconnect with each other for mutual benefit? *The Times*, June 22nd, p. 9.
King, S. (1999), Knowing Your Customer. A Research Report into Customer Information, KPMG Consulting, London.
Krauss, M. (2009), How Redbox is changing retail, *Marketing News*, New York, November 15th, pp. 5–6.
Lee, L. and Ariely, D. (2006), Shopping goals, goal concreteness, and conditional coupons, *Journal of Consumer Research*, Vol. 33, No. 1, pp. 60–67.
Lovinson, J. (2010), Mobile marketing, *Advertising Age*, New York, February 22nd, pp. 58–60.
Murray, C., Talukdar, D. and Gosavi, A. (2010), Joint optimization of product, price, display orientation and shelf-space allocation in retail category management, *Journal of Retailing*, Vol. 86, No. 2, pp. 125–36.
Neslin, S.A. and Shankar, V. (2009), Key issues in multichannel customer management: current knowledge and future directions, *Journal of Interactive Marketing*, Vol. 23, No. 1, pp. 70–81.
Parasuraman, A., Zeithaml, V.A. and Berry, L.L. (1985), A conceptual model of service quality and its implications for future research, *Journal of Marketing*, Vol. 49, No. 4, pp. 41–50.
Parasuraman A., Zeithaml, V.A. and Berry, L.L. (1988), SERVQUAL: Multiple-item scale for measuring consumer perception of service quality, *Journal of Retailing*, Vol. 64, No. 1, pp. 12–40.
Parasuraman A., Zeithaml, V.A. and Berry, L.L. (1991), Refinement and reassessment of the SERVQUAL Scale, *Journal of Retailing*, Vol. 67, No. 4, pp. 420–450.

Rohwedder, C. and Johnson, K. (2008), Pace-setting Zara seeks more speed to fight its rising cheap-chic rivals, *The Wall Street Journal*, February 18th, p. B1–B10.

Sasser, W.E. (1976) Matching supply and demand in service industries, *Harvard Business Review*, November/December, pp. 133–140.

Scanlon, J. (2009), How Kiva robots help Zappos and Walgreens, *BusinessWeek*, April 15th, pp. 3–4.

Shankar, V., Venkatesh, A., Hofacker, C. and Naik, P. (2010), Mobile marketing in the retailing environment: current insights and future research avenues, *Journal of Interactive Marketing*, Vol. 24, No. 2, pp. 111–120.

Sorescu, A., Frambach, R., Singh, J., Rangaswamy, A. and Bridges, C. (2011), Innovations in retail business models, *Journal of Retailing*, Vol. 87, No. 1, pp. S3–S16.

Stross, D. (2010), Why bricks and clicks don't always mix, *New York Times*, September 18th, p. B4.

Sullivan, E.A. (2008), Austerity marketing, *Marketing News*, London, October 15th, pp. 13–14.

Teas, R.K. (1994), Expectations as a comparison standard in measuring service quality: an assessment of a reassessment, *Journal of Marketing*, Vol. 58, No. 1, pp. 132–139.

Teece, D.J. (2010), Alfred Chandler and 'capabilities' theories of strategy and management, *Industrial & Corporate Change*, Vol. 19, No. 2, pp. 297–316.

Useem, J. (2007), Apple: America's best e-etailer, *Fortune*, New York, March 8th, pp. 8–9.

Winstanley, M. (1997), What drives customer satisfaction in commercial banking? *Commercial Lending Review*, Vol. 12, No. 3, pp. 36–42.

Zeithaml, V.A. and J.M. Bitner (1996), *Services Marketing*, McGraw-Hill, New York.

Zott, C. and Amit, R. (2010), Business model design: an activity system perspective, *Long Range Planning*, Vol. 43, No. 2/3, pp. 216–224.

12
The B2B Sector

Recent trends

Over the last four decades many Western firms have become enamoured with exploiting the opportunities for reducing costs by utilising suppliers based in countries such as China or Thailand. Initially these suppliers merely produced standard components but, as their skill levels increased, they moved into contract manufacturing and assembling finished products. Eventually some Western firms relocated their entire manufacturing operations overseas and inevitably also transferred Intellectual Property (IP) to these locations. Preventing IP leaving the factory inside the heads of local staff is impossible but Western firms have accelerated the rate of IP losses by also opening overseas R&D centres. Once this occurred it was inevitable that overseas competitors in the developing nations would soon produce lower priced equivalent products to be sold in both domestic and overseas markets. As the depth of austerity has deepened, Western firms are facing increasing levels of price competition for which they have no defence; and for some firms the inevitable outcome of having trained their new overseas competitors is bankruptcy.

Recently some Western manufacturers have awoken to the risks associated with training of potential competitors in emerging economies. This realisation has been accompanied by recognition of the need to upgrade productivity in their domestic operations in order to counter price-based competition from overseas competitors. Actions to upgrade productivity involve adopting new production techniques, such as lean manufacturing, and investing in process automation. There is also greater recognition that when ongoing rises in energy costs are added to the price of buying goods from overseas suppliers, the issues of shipping costs, customs duty, delivery time and the risk of owning excessive levels of potentially obsolete items all provide a strong economic argument for returning to purchasing from domestic sources (Brown, 2010). Despite this benefit, B2B companies must nevertheless accept that technology in mature industries will become commoditised by the transfer of knowledge to the world's lower labour cost, high-growth emerging economies. Hence Western B2B organisations cannot base their survival strategy on exploiting price-based competition permitted merely by being located nearer to customers in their domestic markets. This is because, in a period of austerity, overseas competitors will increasingly seek to sell at or below cost in order to absorb the fixed overhead costs associated with operating high fixed-cost manufacturing facilities.

Re-emerging home markets

Case Aims: To illustrate that a return to 'onshoring' is assisting some manufacturers to sustain their share in their domestic markets.

Western manufacturers are finding that the impact of lower priced imports is being blunted by their key customers now returning to onshore procurement (Marsh, Shotter and Brown, 2012). In some cases this reflects rising labour costs in countries such as China and weakness of Western currencies relative to the renmimbi. In the past when important major companies went offshore for products they encountered problems over quality, inflexibility of their overseas suppliers and delivery delays. This is the reason, for example, that Numatic, a UK manufacturer of vacuum cleaners, has returned to local sourcing because this (i) improves the response time for key components and (ii) the company no longer has to worry about having sufficient parts in stock when facing a sudden upswing in orders. Another company Group Rhodes, a UK machinery manufacturer, is for the same reasons purchasing 90 percent of their annual requirements for iron castings from local suppliers versus only 40 percent just three years ago. In addition to speed of response and improved reliability, the other advantage is that when a problem arises with a local supplier Group Rhodes can just make the short trip to that supplier's foundry and discuss how to resolve matters.

Sherwood Electonics, based in Nottinghamshire, UK, makes cable assemblies for computers and railway equipment. Over the past 12 months the company has increased the proportion of cable-related components purchased from UK suppliers. In their case a key influence was rising wage and transportation costs for products sourced in Asia, which had risen by 30 percent over the past 18 months. The firm has also been assisted by the fact that some UK suppliers have in recent years adopted new manufacturing technologies which have significantly enhanced their ability to compete with firms located elsewhere in the world.

In the USA, the world's largest construction equipment manufacturer Caterpillar Inc. is examining the benefits of bringing some of their manufacturing operations back to America (Anon, 2012). The assumption is that this move will reduce the company's global supply chain management costs. This strategy could triple the current capacity of hydraulic excavators now produced by the company in the US, increasing Caterpillar's employment levels in their US operations and leveraging excavator production in Japan to serve increasing demand in Asia. Currently the company only produces two excavator models at a facility in Aurora, Ill., plus wheel loaders, soil and landfill compactors, wheel dozers and components. If onshoring is implemented a new facility would be constructed in Aurora to manufacture two new excavator models and possibly several additional excavator models now produced in Japan and exported to the USA.

Some of the factors which have caused domestic problems for American manufacturers in the past when competing against lowest cost imports from emerging nations include higher labour and healthcare costs, pollution abatement fees and relatively higher business taxes. Nevertheless some manufacturers are still turning to onshoring, using production philosophies such as lean manufacturing to improve their productivity and strengthen the ability to deliver superior customer service (Kemp and Mouranie, 2010). The opportunities offered by lean manufacturing are also causing firms from other nations to establish their manufacturing operations in the USA. One example is a Swiss supplier of measuring instruments which now produces 90 percent of its current products at the company's American facility.

Braun Medical, a world leader in intravenous systems and other medical devices, has expanded their US production over the last few years by focusing upon advanced automation, work force training, lean manufacturing and closer integration with the company's R&D operations. Similarly the massive multi-national General Electric is committed to onshoring more of the company's capabilities to the US, especially in the area of aviation-component manufacturing. Such companies are placing emphasis on a number of key activities to enhance the competitiveness of their onshore operations (Buss, 2009), including:

(1) Investing in innovation and achieving greater operational flexibility.
(2) Supply chain consolidation by building closer links with onshore suppliers.
(3) Investing in lean manufacturing practices and other internal processes such as warehouse automation and computer-based stock decision models.

An example of a company which believes innovation is the foundation stone for successful onshoring is Corning, the world's largest maker of

> ceramics and specialty glasses. The company's belief over the superiority of US knowledge workers is reflected in their decision to invest millions in a new plant in Erwin, New York for making automotive catalytic converters and diesel-particulate filters. Ford Motors, despite their problems during the global banking crisis, has become another exemplar of flexible manufacturing to enhance onshore productivity in the company's American production plants. Ford's in-plant body shops now piece together multiple models at each site, even in mixed or random order, and allow rapid model changeovers. Paint and final assembly operations are equally flexible, which when accompanied by reducing the complexities of model changeovers has allowed the company to both improve productivity and boost manufacturing quality.

Innovation

Germany, UK and the USA all initially built their respective economies by exploiting innovation. In mature industry B2B sectors, Germany is the only one of the three countries which in recent years has managed to retain this capability for innovation, as is demonstrated by the country's leadership in the export of premium priced products such as turbines for power generation and advanced machine tools. With price competition neither a feasible nor advisable option for B2B companies in the UK and USA during a period of prolonged austerity, one solution is to exploit leadership in science and technology to produce advanced technology products which can command a premium price in domestic and overseas markets.

Porter and Rivkin (2000) proposed that the three most likely triggers leading to a move to exploit new technology or enter new markets are:

(1) Change in technology.
(2) Change in customers' need.
(3) Change in government regulations.

In some cases two or more of these triggers may emerge simultaneously. An example of this scenario is the need to reduce greenhouse gas emissions due to the rising costs of hydrocarbon fuels and global warming. Examples of technological responses in relation to both these scenarios are electric vehicles or hydrogen powered jet engines. Since World War II the USA has demonstrated how exploiting new technology benefits can maximise the wealth of a nation. The problem for America and Europe is sustaining this outcome in today's global economy when most modern technologies are based around complex systems for which interdependencies exist in areas such as advanced materials and specialist components. This occurs because, once a country permits key elements of these systems to move offshore, difficulties arise for domestic manufacturers wishing to retain their leadership and future control within their industry (Wen-Bin and Hui-lin, 2010).

Loss of control

Case Aims: To illustrate some of the risks of losing control of the proprietary knowledge required to sustain leadership in a high-technology industry.

Although US firms still retain leadership in the global semiconductor industry, China has become the world's largest exporter of IT equipment and many of the components used in US IT products come from China or other Asian economies (McCormack, 2009). The world's leading microchip producer Intel is American, but nevertheless the country's share of semiconductor production capacity has declined with the US no longer being involved in certain key areas of semiconductor equipment manufacturing. Christensen et al's (2004) concern is that the industry is approaching 'decoupling point', with America becoming unable to retain a leadership position in key industries such as computing and mobile communications devices. Once decoupling occurs opportunities for overseas manufacturers upstream in the supply chain are opened up.

Approximately 80 percent of all 300 mm fabs are manufactured outside America, even though US semiconductor companies still retain a 48 percent global market share and remain undisputed technological leaders. Although these American 'fabless' semiconductor companies have been successful, when a major new technology eventually emerges these firms may not be able to design next generation components without sharing their R&D knowledge and capabilities with the overseas fab foundries. The event could cause a significant proportion of America's high value-added domestic economic activity to disappear overseas (Petrick, 2009; Wolfe, 2009). The business risk is illustrated by global distribution of the components of the Kindle, Amazon's e-reader. The estimated cost of the Kindle is $185 of which only $40–$45 is estimated to be captured within the US economy; the balance being retained by overseas firms based in Asia. Another example is Apple's iPod where the vast majority of the product's value for all the manufactured parts now occurs offshore (Pisano and Shih, 2009).

It can be argued that Western companies such as Apple and Amazon have become highly successful by focusing on design and the downstream marketing of their products. The unanswered question is: what period of time will pass before these overseas subcontractors move into producing their own branded products which are then marketed across the Western world? Based upon prior events such as Japan's Sony and Korea's L.G. success in electronic consumer goods, and Samsung's latest smartphone, there will come a time when companies such as Apple will face intense competition from lower priced 'me too' propositions unless they are able to remain ahead by developing higher performance, next generation products.

Regaining control

Within the IT and related industries, some observers believe that retaining supply chain control involves combining high value technology with high technology support services. This philosophy is exemplified by IBM, which as a company learnt in the 1990s the risks of becoming reliant upon manufacturing computer hardware when overseas companies based in lower cost areas of the world began to acquire sufficient capability to offer lower price clones. IBM's recovery strategy focused upon becoming the provider of high-technology services in the field of complex data management and making this expertise available in the form of combined hardware and software solutions. To achieve this goal IBM developed a very effective vertical integration strategy. Their major research facility, the Watson Research Centre in Westchester, NY, is located near to the company's semiconductor R&D and 300 mm fab manufacturing facility in Fishkill, NY. Also nearby is a large computer systems manufacturing and assembly operation. The co-location of these various facilities provides the synergies required to be a leading company in the field of exploiting concurrent innovative tangible product and service developments (Stanko and Calantone, 2011).

Economic recovery in the West requires moving away from a consumption-led growth strategy based upon excessive consumer borrowing and instead focusing upon innovation, assimilating existing technologies and developing new ones to enhance productivity and dominate export markets in order to rebuild nations' productive assets. Only once this has occurred will per capita incomes begin to increase and the current period of austerity be brought to an end (Wolfe, 2009). Key to this outcome is R&D. Traditionally R&D has been led by Original Equipment Manufacturers (OEMs). More recently this model is being replaced by a 'value stream' model in which R&D is distributed throughout entire supply chains (Petrick, 2009). The success of this latter model requires much more collaboration among the members of a supply chain and greater acceptance of an open innovation philosophy. Concurrently however lead firms must seek to retain control of proprietary areas of technology in order to determine the future source of value within a market system.

Sustained transformation

Case Aims: To illustrate the requirement to be prepared to transform an organisation in the face of changing opportunities.

The speed of change caused by new technology demands that Western B2B organisations retain a fundamental belief in the need for continual transformations, either through technological developments or through acquisitions. Furthermore these organisations must accept that some of these transformations will be central to change in the markets served and the technology utilised (Tebo, 2009).

An excellent example of repeated fundamental transformations is provided by the American company DuPont (Chowdhry, 2011). Having started life as a manufacturer of explosives, when this sector moved into maturity the company underwent transformation into a chemical company. This was followed by having the foresight in 1903 to merge capabilities in chemicals and polymers and invest in in-house R&D laboratories. This decision led to landmark inventions of neoprene and nylon in the 1930s, which provided the basis for the world's new polymer industry. By the 1980s the polymer industry had become increasingly price competitive as other chemical companies built production facilities elsewhere in the world, eventually leading to excess global production capacity.

The next transformation at DuPont was to focus on market-driven science to identify new market opportunities. The company recognised 'scale of speed' in innovation as becoming an increasingly important capability. DuPont has a key advantage over many competitors in that the breadth and depth of the organisation's technological knowledge permits faster development of commercially viable new products. Most recently DuPont's R&D philosophy is being driven by a focus on the three key mega-events of (i) total world population growth, (ii) the increasing size of the world's affluent middle class and (iii) protecting the planet's environment and scarce resources. This has led to investment in biotechnology which the company perceives as critical in the fields of advanced chemicals, pharmaceuticals and agriculture. To achieve its goal, DuPont acquired Pioneer, one of the world's largest seed companies. This has assisted the latest transformation of moving DuPont from being a volume-driven chemicals producer to a value-driven science company.

Low-technology innovation

For low-technology B2B industries there is often little opportunity to utilise new technology as a strategy for defeating price-based competition from overseas, and this problem is being exacerbated by the current prolonged economic downturn. In certain sectors such as the food industry the expense of product distribution, perishability or import restrictions may provide a defence against lower priced competitors. Increasingly however improving global logistics and higher quality control standards in emerging nations do mean fewer and fewer companies in B2B markets can expect to remain unaffected by the threat of lower cost imports. Hence low-technology firms based in developed economies need to indentify a strategic defence based upon innovation.

Two entrepreneurial options are product innovation to upgrade the performance of finished products or process innovation to reduce production costs, enhance production yields, increase production volumes or fulfil environmental or energy conservation targets (Markus, Bergfors and Larsson, 2009).

Examples of low-technology mature industries where innovation may be limited to process innovation are food and drinks, clothing, footwear and textiles (Zahra et al, 2006). In these sectors, Muscio, Nardone and Dottore (2010) proposed, there exist the following demand typologies:

(1) Real demand where firms can improve their products/processes.
(2) Latent demand where firms can exploit such needs through innovation.
(3) Potential demand innovation where needs are not yet realised because no firms are capable of responding to the identified innovation challenge.

These researchers examined the Italian food industry and concluded that, although occasionally firms may be able to exploit innovation to offer the end user an enhanced eating experience or improved product convenience, these opportunities are relatively rare. In contrast process innovation offers a range of opportunities, including reducing environmental impact, improving food safety, lowering consumption of non-sustainable resources such as energy or water and increasing production yields and productivity. In the case of functional foods, the changing attitude of consumers towards healthier eating has necessitated a focus on process innovation to incorporate raw materials that lead to the production of finished goods offering benefits such as fewer calories or reduced fat levels. The search for a higher food safety is most frequently encountered in product categories such as wine and preserved vegetables where fermentation may cause the undesired proliferation of different sorts of toxins and pathogens. The problem is that customers expect ever higher food safety standards but are usually unwilling to pay a premium price to gain access to such products. Meeting intermediaries' demands for longer shelf life has made processing more expensive but again recovery of these higher costs through raising prices is rarely an achievable aim. Hence food firms usually have to focus upon innovation to reduce costs in other areas of their operations such as being more efficient in their use of energy and water. The increasing complexity of introducing new technology and the highly integrated nature of market systems does mean that many low-technology manufacturers, such as the giant branded consumer goods company Proctor & Gamble, have moved towards open innovation as a mechanism to sustain revenue in mature markets impacted by a long period of austerity. The obstacle facing many firms wishing to adopt this approach however is a corporate culture characterised by a 'not-invented-here' attitude. In commenting upon this Lichtenthaler et al (2011) concluded that open innovation cultures are of four different types:

(1) *Technology isolationists* are characterised by high levels of both not-invented-here and not-sold-here attitudes.
(2) *Technology fountains* are companies which are sources of technological knowledge that they partly transfer to other companies, but are still characterised by a strong not-invented-here attitude.
(3) *Technology sponges* actively absorb external technology, but do not transfer their own technology to external partners.
(4) *Technology brokers* pursue both inbound and outbound open innovation without concerns about not-invented-here activities.

The lowest average return on sales was found among the technology fountains who focus on internal technology development. Technology isolationists who rarely engage in open innovation had an average return on sales. Technology sponges and technology brokers achieve an above average return on sales. Given the superior financial performance associated with some types of open innovation, Lichtenthaler et al proposed that there are a number of actions for reducing the not-invented-here and not-sold-here tendencies. These include:

(1) Managers communicating the importance of adopting an open innovation strategy.
(2) Top management actively supporting open innovation.
(3) Establishing incentive systems to reward open innovation successes.
(4) Designing organisational structures to optimise the use of open innovation.
(5) Embedding open innovation into the corporate culture.

Responding to the threat of low cost competition

Case Aims: To illustrate how a mature conservative sector of manufacturing is beginning to respond to the threat of low cost competition.

Increasing global competition has adversely impacted the US household furniture manufacturing industry with the non-upholstered wood household furniture manufacturers losing most market share to low cost imports from the emerging nations. The decline in furniture manufacturing has also greatly affected the US hardwood lumber industry. In contrast other sectors of the US furniture and related product manufacturing have been able to defend their competitive position (Buehlmann and Schuler, 2009).

One of the reasons for the poor performance of the US non-upholstered wood household furniture industry is the sector's reputation for being more conservative and significantly less innovative than other manufacturers. Other influencing factors include intensive domestic competition eroding margins, thereby lowering the affordability of investing in new technology, the relative simplicity of the sector's manufacturing processes and the nature of the retail market system in the USA. Non-upholstered wood household furniture is mostly sold as standardised mass-produced products. In contrast upholstered household furniture involves the customer selecting the fabric to be used. As a consequence this latter type of furniture cannot be produced until the customer has placed an order. Wait times then become critical and, with air transportation of the product from an offshore location being extremely expensive, domestic manufacturers are in a much better position to service this sector of the market.

The steady rise in labour costs in many emerging nations, coupled with higher transportation costs, political unrest and time in transit is beginning to erode the benefits of locating wood furniture manufacturing operations in offshore locations. As a consequence domestic manufacturers are revisiting their strategies over plant locations, perceiving competitive advantages from capacity evolving toward a more localised domestic location approach to manufacturing and supply chain management.

Successful wood furniture manufacturers are forging closer links with intermediaries to more rapidly identify changing consumer needs, provide superior supply chain services and exploit new technology to offer customised products (Lihra, Buehlmann and Beauregard, 2005). These changes are causing manufacturers to outsource more subcomponent work to specialised entities, whilst they focus on managing more efficient supply chains, upgrading their assembly capability and distribution operations. These manufacturers are also in a much better position to respond to increasing sales channel diversity in the USA as sales have shifted from conventional household furniture retailers to mass market retailers such as Wal-Mart and Target, department stores such as JC Penney and manufacturer-owned stores like IKEA and Ethan Allen. All of these trends mean that US wood furniture manufacturers and their suppliers should introduce operations philosophies similar to the car industry, with success being determined by improved marketing practices, supplying high quality products tailored to the consumers' needs and investing in new technology to support more efficient production processes at a competitive price (Vlosky, Ozanne and Fontenot, 1999).

For Western manufacturers of low cost, high volume standardised goods survival by using innovation to move up-market offering a technologically superior, premium priced niche product is rarely a feasible option. Usually the only realistic survival strategy is to find ways of matching overseas competitors' production costs in their own domestic manufacturing operations. An example of a company implementing this strategy is Amtico, a UK manufacturer of vinyl floor tiles (Brown, 2012). The CEO of the company, Jonathan Duck, concluded that in order to survive his UK factories needed to overcome the 40 percent adverse cost gap versus Chinese producers. By focusing upon running the factories more efficiently, re-designing work flows and reducing product changeover times by 2012 the company has been able to reduce costs such that the factory output is only 5 percent more expensive than overseas competitors. In addition the organisation developed a lower cost more versatile floor tile which is marketed alongside the company's primary Amtico range. The other strategic benefit is company location. Most B2B customers such as hotels and department stores postpone decisions about flooring until the very last point in a refurbishment or new build project and then want virtually immediate delivery. By being located close to customers Amtico is in a much better position to respond to these order patterns and can react more rapidly to changes in demand than the company's overseas competitors.

Going green

Some politicians in Western nations are currently articulating the view that there is a need to increase the size of their nation's manufacturing sector. In a period of prolonged austerity this is unlikely to occur unless firms are able to establish a competitive advantage which protects their operations from price-based overseas competition. To achieve this goal Western manufacturers require an ability to exploit superiority in technological innovation as the basis for developing products which more adequately meet the needs of customers than lower priced offerings from overseas suppliers.

One area of opportunity which has received extensive coverage in the literature for a number of years is for Western firms to exploit technology as the basis for making available greener products. Dangelico and Pujari (2010) proposed that radical green product innovation could be divided into two types; namely the use of new technologies (e.g. hybrid vehicles) or the replacement of one critical component with another that reduces the overall environmental impact of the product. They contrast these solutions with incremental green innovations. These involve the use of existing key product dimensions to improve eco-efficiency by replacing conventional materials with materials that have less environmental impact or product designs that permit recycling of products at the end of their useful lives.

The problem with green innovation is that in many cases environmentally responsible products are often more expensive to produce and most customers in periods of austerity tend to place greater emphasis on affordability than exhibiting concerns about protecting the environment. In B2B markets such attitudes only tend to change when the customer is facing the risk of incurring heavy financial penalties as a consequence of government legislation. In such instances Western manufacturing firms are beneficiaries of government actions in those cases where overseas competitors in the emerging economies lack the technological capability to produce the products required to meet higher environmental standards.

Opportunities for green innovation have begun to change. This is because the ongoing rise in the cost of oil is causing B2B customers to seek suppliers capable of assisting them to achieve the goal of reducing energy consumption in their internal activities and also to develop the more energy efficient products being demanded by customers. These pressures are very apparent in the airline industry where the rising cost of jet fuel has caused carriers to pressure companies such as Boeing and Airbus Industries to develop more fuel efficient aircraft. In turn these manufacturers are demanding suppliers of engines and materials used in aircraft construction develop new products capable of contributing to reducing fuel consumption in their next generation of aircraft.

Over the last decade the enthusiasm for green technology led to a proliferation of new ventures. Following the onset of the global recession in 2007/8 many of these were forced to shut down. This outcome reflected a lack of market demand as customers became more price conscious. Day and Schoemaker (2011) concluded that survival in the face of a long-term period of economic uncertainty requires green technology companies to develop long-term staying power. Additionally firms must understand the available opportunities and how to remain ahead of the competition. These researchers offered the following guidance, which they evolved from analysing success and failure across a broad array of emerging technologies, including the biosciences, nano-manufacturing, advanced materials, space science and information technology:

(1) *Timing is everything*, reflecting the fact that success in green technology can only occur when customers perceive the need to choose products which offer green attributes.
(2) *Change occurs slowly*, meaning early entrants into a new green market may have to wait many years before there is a sufficiently large critical mass of customers to provide the basis for a financially viable business.
(3) *Collaboration is crucial* because in many cases no single company will have sufficient influence to persuade customers or governments that change is necessary.
(4) *Innovative solutions are multiple*, which means innovators face both opportunities and threats. An example is the Telsa sports car developed in part by individuals from the IT industry exploiting their understanding of lithium batteries. The threat is that this apparently viable solution may be overtaken by a superior solution, such as advances in the use of hydrogen power cells under development by BMW and Toyota. Another example is apparent in the sustainable energy industry as the financial viability of wind power has been usurped by ongoing advances in solar energy technology.
(5) *Keep options open* during the development and launch of a new product because it can be the case that new technology may emerge that offers a superior solution.
(6) *Identify revenue flows* because, should these undergo change, an apparently viable solution may later face problems. This outcome has occurred in the solar energy market where initially many governments were offering subsidies that made installing solar panels an attractive proposition. More recently governments facing public sector deficits have either reduced or removed these subsidies, leaving the solar energy industry less able to sustain claims over the financial benefits of their technology.
(7) *Think outside the box*, because this can result in the identification of completely different ways of solving a problem. This is exemplified in the energy market where genetically engineered plants and micro-organisms are now being used in fermentation to concert biomass such as converting switchgrass into jet fuel.
(8) *Exploit the benefits of collaboration*, because this can assist in spreading the risk whilst concurrently providing access to new technology from sources outside of the organisation.
(9) *Watch for weak signals*, because this can provide earlier warnings of a potential threat or assist in identifying a solution to a technological problem which currently seems un-resolvable.

(10) *Remain flexible*, because in this way the organisation can more rapidly change direction if an emergent technology appears to offer a more effective solution or changing market circumstances require a complete rethink of the organisation's future business strategy.

Protectable competitive advantage

Case Aims: To illustrate how specialist capability can permit Western manufacturers to remain ahead of the competition.

One survival strategy is to develop unique competences which cannot be easily duplicated. For some Western manufacturers one way of achieving this is to combine high-level craft skills evolved over many years with the latest high-technology production equipment (Marsh, 2012). One firm using this approach is Keeler, a world-leading UK manufacturer of optical instruments. The company has the capability of producing small quantities of highly specialised lenses for use in microscopes. Keeler is a subsidiary of Halma, which, in addition to their optical equipment operation, also has operations in areas such as specialist devices for preventing explosions in processing vessels operated at high pressure.

Another UK firm exhibiting a similar strategic positioning is Spectris, manufacturer of a range of instruments and data analysis machines for environmental monitoring and control systems for chemical plants. This company has sales in excess of £1 billion and employs approximately 6,000 people. Three more firms with the same market positioning are Renishaw, Spirax Sarco and Rotork. These companies are respectively the biggest in the world in the highly specialist sectors of measuring probes for machines tools, steam control systems for boilers and value actuators for controlling fluid flows in chemical and oil plants.

A common trait of these firms is that much of their output is purchased out of their customers' operating budgets. As a result these firms are relatively immune to cutbacks in capital expenditure which occur during periods of austerity. Another austerity-proofing factor is that these companies' products assist their customers either to conform to new environmental regulations or reduce energy consumption in large scale processing operations. Sustainable sales revenue is also aided in these companies by employing specialist, highly trained engineers who in many cases have greater knowledge than their customers of specific process technologies within the latter's production operations. As a consequence the customers are highly reliant upon these suppliers as a source of knowledge. This in-depth knowledge assists these firms in identifying new product opportunities within their customers' operations or in becoming aware ahead of the competition of their customers' plans for expanding or changing their future operations.

B2B services

The knowledge worker superiority of Western nations in professions such as accountancy, banking and law has permitted firms in these sectors to sustain their global dominance in B2B markets despite the onset of economic austerity. During the current economic downturn these superior knowledge firms have been able to sustain revenue by expanding their operations in emerging economies such as India and China. In the case of legal practices, this sector is greatly assisted by the fact that most international law is based upon the British or American legal system, thereby providing an institutional framework common to commercial transactions in most countries around the world (Pinnington and Gray, 2007). Even though strong regulatory differences exist between countries, some large corporate law firms have successfully implemented global strategies by retaining their offices in the UK or USA and then acquiring smaller local practices where an in-country presence and local knowledge is necessary to meet client needs (Segal-Horn and Dean, 2011).

The performance of the major Western banks over the last few years has been drastically impaired by problems associated with the sub-prime mortgage crisis in the USA, involvement in the sovereign debt problems in the EU and a decline in M&A activity (Jegher, 2010). Nevertheless the Western banks still retain a superiority of knowledge in relation to corporate international financial transactions and the international lending markets. As a consequence, similar to law firms, some compensation for a downturn in revenue from their domestic operations has been achieved by expanding their operations in emerging markets.

Possibly the greatest beneficiaries of the growth in world demand for services have been the British and American accountancy firms. Initial global expansion occurred when Western multinationals demanded their auditors accompany them overseas in order to ensure financial matters were being correctly managed in their international subsidiaries (Laird, Kirsch and Evans, 2003). The expertise which these firms gained in international financial management, when linked with their understanding of financial matters associated with fulfilling stock market regulations on Wall Street or in the City of London, resulted in them becoming the favoured service providers when emerging nation firms need to upgrade the quality of their audit operations or engage in international financial transactions. Electronic data storage, sophisticated computer-based information management and the Internet have also greatly assisted Western professional services providers to meet customer needs in international markets. Initially this capability was exploited to enhance service delivery to their domestic clients. Subsequently this acquired expertise in electronic data management proved invaluable as the professional services sector has moved into other nations across the globe.

Western manufacturing firms have also come to recognise that significant new opportunities are created by exploiting their IT capabilities to expand the range of services that can be made available to their clients. Similar to the professional services sector, Western manufacturers' expertise in electronic data management has permitted expansion of their services provision activities to compensate for declining sales for tangible goods following the onset of economic austerity. Some of the world's leading manufacturers such as GE Corporation and Rolls-Royce are currently prepared to accept relatively low prices from the sale of their products because after-sale services provide a long-term source of sustained profitability (Oliva and Kallenberg, 2003). The benefit to customers in purchasing both products and services is enhanced productivity, because suppliers can deliver services such as equipment maintenance, resolve complex operating problems should they arise and offer expertise to identify new ways of improving operational efficiencies (Gebauer, 2007).

Ulaga and Reinartz (2011) identified four sources of competitive advantage for manufacturers offering combined product and service packages, namely:

(1) *Installed base product usage and process data* exploiting installed products' provision of unique data sources, which can be utilised by the supplier to offer customers ongoing support and assistance. This occurs because many products in B2B markets are equipped with information and communication technologies that permit the supplier to monitor the product in use at a customer's location. For example, one forklift trucks manufacturer remotely monitor operations at customer locations on a real-time, 24/7 basis. This activity provides data on how many hours the forklift truck runs per day, how many hours of downtime the equipment endures and permits the manufacturer to offer guidance on issues such as servicing and repairs.
(2) *Product development advantages*, because by monitoring their products in use manufacturers can identify problems and opportunities, which provide the basis for new product ideas. An example is provided by a tyre manufacturer who used acquired knowledge to develop a new tyre casing that allowed re-grooving and retreading of tyres, enhancing the number of miles customers could obtain from one set of tyres.
(3) *Service-related data processing* involving remote monitoring of products in use at the customer's location and providing early warning of potential problems, which can be avoided by instructing customers to initiate service cycles or making adjustments to how the products are being utilised.
(4) *Risk assessment*, whereby those manufacturers who monitor can identify early warnings of potential problems such as equipment breakdown and this permits the manufacturer to guide customers on how to avoid an emerging risk.

Ulaga and Reinartz (2011) described the approach of supplying both products and services as 'hybrid marketing' and concluded that the concept requires a very different approach to that of merely selling tangible goods. The sales process is more complex, usually involving lengthier negotiations focused on meeting customer-defined specifications. With hybrid offerings, the proposition is usually much less well defined at the beginning of the negotiation process. Strong customer involvement is needed to reach a final purchase specification. Closing the sales often involves multiple actors from both the customer and the supplier organisations.

Hybrid services

Case Aims: To illustrate how aerospace manufacturers now provide significantly more than just tangible products.

Exemplars in the provision of hybrid products are provided by the aircraft industry where post-purchase product support does not emphasise just maintenance but also proactive responsiveness in assisting airlines optimise fleet operations (Larson, 2008). A leader in the field is Rolls-Royce, which through their CorporateCare programme provides service coverage that extends to scheduled and unscheduled repair; parts replacements, engine removal and reinstallation; engine transportation expenses and engine leasing services. Regularly up-dated service bulletins covering mandatory or recommended activities are included in Rolls-Royce Engine Management Programme publications. The company also provides training and educational up-dating services to customers' employees. Monitoring of engine performance when aircraft are flying is achieved by using a Web-enabled real-time engine health viewing system. CorporateCare is transferable from operator to operator and usage hour agreements cover all expenses for scheduled and unscheduled requirements, engine maintenance, replacement parts and parts replacements from 250 Rolls-Royce authorised maintenance centres at strategic locations across the world.

Hawker Beechcraft Support Plus programme allows owners to select parts-only coverage or a programme providing comprehensive maintenance. Coverage for parts includes scheduled and unscheduled inspections, maintenance and component removals. The more comprehensive plan covers labour costs for the same items when purchased through Hawker Beechcraft's global network of authorised service centres. Bombardier Aerospace has created their Classic support programme for the company's Learjets. The programme recognises that operators of older aircraft have very different needs than those of newer generation aircraft. The programme provides full engine maintenance and technical support. The company has established two Customer Response Centres, one for Learjet and another for their Challenger and Global aircraft, which provide single-point-of-contact assistance to customers on a 24/7 basis. The company's PartsExpress support is responsible for dispatching parts, technical support and co-ordinating the activities of the company's specialist mobile repair teams.

References

Anon (2012), Caterpillar shifts from offshoring to onshoring back to the U.S., *Material Handling & Logistics*, March 12th, 3–4.
Brown, A.S. (2010), Manufacturing at the crossroads, *Mechanical Engineering*, Vol. 132, No. 6, pp. 30–40.
Brown. (2012), J.S. Floor tiles chief shows true grout to take on rivals, *Financial Times*, London, February 17th, p. 20.
Buehlmann, A. and Schuler, A. (2009), The U.S. household furniture industry: status and opportunities, *Forest Products Journal*, Vol. 59, No. 9, pp. 20–28.
Buss, D. (2009), Rethinking manufacturing strategy, *Chief Executive*, New York, August, pp. 28–33.
Christensen, C., Musso, C. and Anthony, S. (2004), Maximizing the returns from research, *Research–Technology Management*, Vol. 47, pp. 12–18.
Chowdhry, U. (2011). Transforming American industry, *Research Technology Management*, Vol. 50, No. 2, pp. 18–27.
Dangelico, R.M and Pujari, D. (2010), Mainstreaming green product innovation: why and how companies integrate environmental sustainability, *Journal of Business Ethics*, Vol. 95, pp. 471–486.
Day, G.S. and Schoemaker, P.J.H. (2011), Innovating in uncertain markets: 10 lessons for green technologies, *Sloan Management Review*, Vol. 52, No. 4, pp. 38–46.
Gebauer, H. (2007), An investigation of antecedents for the development of customer support services in manufacturing companies, *Journal of Business-to-Business Marketing*, Vol. 14, No. 3, pp. 59–70.
Kemp, J.W. and Mouranie, C. (2010), Manufacturing can be competitive in the U.S., *Northeast Pennsylvania Business Journal*, Vol. 25, No. 7, pp. 6–9.
Laird, K.R., Kirsch, R.J. and Evans, T.G. (2003), A marketing resource-based model of international market entry and expansion for professional services, *Services Marketing Quarterly*, Vol. 24, No. 4, pp. 1–13.
Larson, G.C. (2008), Maintenance cost control programs, *Business & Commercial Aviation*, Vol. 103, No. 2, pp. 128–138.
Lichtenthaler, U., Hoegl, M. and Muethel, M. (2011), Is your company ready for open innovation?, *Sloan Management Review*, Vol. 53, No. 1, pp. 45–54.
Lihra, T., Buehlmann, U. and Beauregard, R. (2005), Mass customization of wood furniture as a competitive strategy, *International Journal of Mass Customization*, Vol. 2, No. 3/4, pp. 200–215.
Jegher, J. (2010), Top corporate banking trends 2010, *Commercial Lending Review*, July–August, pp. 48–53.

Porter, M.E. and Rivkin, J.W. (2000), Industry transformation, *Harvard Business Review*, July/August, pp. 10–18.
Markus, E., Bergfors, A. and Larsson, D. (2009), Product and process innovation in process industry: a new perspective on development, *Journal of Strategy and Management*, Vol. 2, No. 3, pp. 261–276.
Marsh, P. (2012), Niche spells success for manufacturers in the UK, *Financial Times*, London, February 7th, p. 23.
Marsh, P., Shotter, J. and Brown, J.M. (2012), Industry switches to Britain, *Financial Times*, London, February 7th, p. 4.
McCormack, R. (2009), The plight of American manufacturing, in *Manufacturing a Better Future for America*, Alliance for American Manufacturing, Washington, DC, pp. 67–76.
Muscio, A., Nardone, G. and Dottore, A. (2010), Understanding demand for innovation in the food industry, *Measuring Business Excellence*, Vol. 14, No. 4, pp. 35–48.
Oliva, R. and Kallenberg, R. (2003), Managing the transition from products to services, *International Journal of Service Industry Management*, Vol. 14, No 2, pp. 160–172.
Petrick, I. (2009), Supply chain globalization: How surviving SMEs can position themselves for the future, in *Manufacturing a Better Future for America*, Alliance for American Manufacturing, Washington, DC, pp. 43–54.
Pinnington, A.A. and Gray, J.T. (2007), The global restructuring of legal services work? A study of the internationalisation of Australian law firms, *International Journal of the Legal Profession*, Vol. 14, No. 2, pp. 148–170.
Pisano, G. and Shih, W. (2009), Restoring American competitiveness, *Harvard Business Review*, June/July, pp. 3–14.
Segal-Horn, S. and Dean, A. (2011), The rise of super-elite law firms: towards global strategies, *The Service Industries Journal*, Vol. 31, No. 2, pp. 195–213.
Stanko, M.A. and Calantone, R.J. (2011), Controversy in innovation outsourcing research: review, synthesis and future directions, *R&D Management*, Vol. 41, No. 1, pp 8–20.
Tebo, P. (2009), Building business value through sustainable growth, *Research–Technology Management*, Vol. 48, No. 5, pp. 28–32.
Ulaga, W. and Reinartz, W.J. (2011), Hybrid offerings: How manufacturing firms combine goods and services successfully, *Journal of Marketing*, Vol. 75, pp. 5–23.
Vlosky, R.P., Ozanne, L.K. and Fontenot, R.J. (1999), A conceptual model of U.S. consumer willingness to pay for environmentally certified wood products, *Journal of Consumer Marketing*, Vol. 16, No. 2, pp. 12–14.
Wen-Bin, C. and Hui-lin, L. (2010), Interdependence between overseas and domestic R&D activities, *Asian Economic Journal*, Vol. 24, No. 4, pp. 305–332.
Wolfe, R. (2009), U.S. business R&D expenditures increase in 2007; Small companies perform 19% of nation's business R&D, *InfoBrief* (NSF 09-316), National Science Foundation, Washington, DC.
Zahra, S.A., Sapienza, H.J. and Davidsson, P. (2006), Entrepreneurship and dynamic capabilities: a review, model and research agenda, *Journal of Management Studies*, Vol. 43, No. 4, pp. 917–955.

13
Small Firms

Sector importance

By the 1980s, the small and medium size enterprise (SME) sector in both the USA and Western Europe had become an increasingly important source of employment and a significant contributor to Gross Domestic Product (GDP). In the UK by the end of the 20th century the small firm sector, which is defined as all firms with up to 249 employees, was providing 55 percent of all employment and contributing over a quarter of the nation's total GDP. Elsewhere within the European Union (EU), small firms have an even more important economic role, generating over 65 percent of total employment.

Most Western governments now recognise the importance of the SME sector as a source of new jobs and many governments have invested heavily in schemes aimed at increasing the rate of small business creation. Whether these initiatives have achieved the desired outcome of increased job creation is however a somewhat contentious issue. Analysis of the impact these support initiatives have had in the UK, for example, indicates that over the long term they have a negative impact on employment levels (Storey, 2002).

Most small firms operate in market sectors where it is difficult to be different, supply may exceed demand, competition is intense and market entry by new unskilled people is not uncommon. Under these circumstances, profitability tends to be very low and opportunities to generate a scale of profit sufficient to fund business expansion are virtually non-existent. Scenarios of this nature are to be found in abundance in highly fragmented service sectors such as retailing, hairdressing, hotels and restaurants. The survival of this type of firm depends upon high consumer confidence and rising consumer incomes and hence during an economic downturn small business failure rates increase dramatically. Based upon data from both the UK and the USA this outcome is currently being reflected by the increasing number of empty small independent retail outlets in towns and cities, accompanied by a steadily rising increase in small firm closures. Most small firms rely on their local bank as a source of external borrowing and further pressure on small business in the current downturn is banks unwilling to lending to smaller organisations who are perceived as a high risk proposition (Nash and Zeuli, 2011). A 2011 Kaufman Institute survey of small firms in the USA reported 89 percent of small businesses that applied unsuccessfully for bank loans felt that tighter lending restrictions were to blame. The lack of access to capital to grow these businesses was also a likely source of reduced or lost sales forcing many businesses to default on the existing loans which in turn has often led to bankruptcy (Jeffries, 2011).

Survival

The longer the duration of the current period of austerity the greater are the risks of business failure among small firms. Firms most likely to survive are those who over the years through offering superior service, quality or value for money have built a large following of loyal customers. This is because when consumers are being more careful over their expenditure decisions, they tend to opt for those suppliers who have consistently provided high purchase satisfaction. Those small firms which rely upon lower prices and have a limited base of loyal customers are the most likely to fail (Saridakis et al, 2008).

Where revenue is declining one solution is to focus on those customer groups that are growing and/or whose spending power is less likely to be adversely affected by an economic downturn. One opportunity which exists is to exploit population ageing by focusing upon the provision of services for older people in higher socio-economic groups. An alternative strategy is repositioning. This will usually involve some degree of business restructuring in order to provide new services to those seeking more affordable propositions or moving up-market offering services to a wealthier customer base.

Low tech innovation

Case Aims: To illustrate common innovation practices in a low-technology service sector.

Firms operating in low-technology service sectors can expect to face greater difficulties in identifying innovative actions that can distinguish them from their competitors located in the same vicinity (Petrou and Daskalopoulos, 2009). To assess the impact of the current economic downturn, Palmer and Griswold (2011) examined the behaviour of small restaurants in Illinois, USA. They found that although virtually all the business owners indicated that food and service quality were more important sources of competitive advantage than low prices, all owners reported that the downturn has prompted cost-related adaptations

> to their menus. This has involved adding menu items with lower cost ingredients or raising menu prices for unchanged, popular menu items and the introduction of loss leader menu specials.
> In making menu changes, only a minority of owners were seeking to attract new customers. The majority had introduced menu or service amenity changes to maintain or increase current customer visits. Most of the restaurants did not perceive their innovation was either 'cutting edge' or radical, believing any proactive innovation was more likely to occur among the national chain food franchises. Most of the reported significant operational changes only occurred following innovative moves by the national franchises.

Market understanding

Narver and Slater (1990) proposed that market orientation consists of three elements, namely customer orientation, competitor orientation and interfunctional coordination. In their view, continuous innovation is implicit in the successful exploitation of these elements in order to sustain long-term profitability. Kohli and Jaworski (1990) suggested the concept of market intelligence rather than customer focus is the central element of market orientation. This is because market intelligence is a much wider concept than customer focus, covering variables such as exogenous market factors that may affect customer needs and preferences both currently and in the future. To optimise performance firms should not exhibit a reactive response to changes in market environments, but instead need to develop a proactive orientation that permits them to remain ahead of the competition.

Few small firms have either the skills or the financial resources to implement large-scale market research studies. The issue arises therefore of whether the concept of market orientation and customer intelligence is relevant in the small firms sector and can help organisations achieve a higher level of innovativeness. One study which appears to confirm this outcome was carried out by Verhees and Meulenberg (2004). They researched small firm behaviour in a low-technology sector, small commercial rose growers in Holland, where customer demand is relatively homogeneous. A positive relationship was identified between innovativeness and the acquisition of customer market intelligence and market knowledge provided by the firms' suppliers. Higher levels of innovativeness were found to be associated with producers offering a broader and more frequently revised product range which in turn sustained charging higher prices.

One of the debates in the entrepreneurship literature is the apparent contradiction that although firms are advised to use customer data to direct their innovation programmes, some of the most successful entrepreneurs appear to ignore such guidance. There are at least two reasons to explain their behaviour. Firstly in many cases the potential customer does not have the knowledge or understanding to express a desire to purchase a radically new product. Secondly successful entrepreneurs do appear to have this innate ability to intuitively identify future market opportunities and hence require little or no input from either others within their industry or from potential customers (Chaston, 2010).

Non-use of customer intelligence led Atuahene-Gima (1996) to conclude that customer information might actually reduce the degree of newness that a small firm is able to achieve. This is because the input of customer data may cause the innovator to take a less radical approach during the product development phase. In an attempt to validate this concept, Verhees and Meulenberg (2004) measured what they described as 'domain-specific innovativeness' which relates to the willingness of the business owner to learn about and exploit innovations in a specific domain. They concluded that owners/managers who exhibit a high level of domain-specific innovativeness are less market orientated and produce more radical new products. In contrast market-orientated firms tend to be less innovative, thereby suggesting that input from customers may result in a more conservative approach reflecting a desire to satisfy customers' most immediate and obvious needs.

External knowledge

Small firms engaging in innovation face certain disadvantages compared to their large firm counterparts. Firstly large firms may have stronger cash flows to fund R&D or can engage in external borrowing. Secondly large firms can often access a wider range of knowledge sources. Thirdly R&D is a higher relative cost to small firms because (i) they have to invest a higher share of their total resources in R&D, (ii) they face possible high entry costs to create new R&D facilities, (iii) R&D cost recovery has to be spread across a smaller sales base, (iv) small companies face problems accessing external loans to fund R&D and (v) unlike large firms, small organisations cannot spread risk by managing a portfolio of different R&D projects.

Rammer et al (2009) concluded that many small German firms operate without any in-house R&D facilities. Instead R&D is undertaken on an occasional basis, often in response to requests from departments such as production or marketing. A major drawback is that this type of R&D is often less sophisticated in terms of technological outcomes. Furthermore the lack of a permanent R&D facility limits the firm's ability to continuously monitor relevant technology trends and reduces the capacity to acquire new external knowledge. The majority of innovative firms without permanent in-house R&D that Rammer et al have studied refrain from undertaking any kind of leading edge R&D, relying upon technology and knowledge inputs from external sources such as suppliers or trade associations. Systems for exploiting internal knowledge and accessing external sources do exist however in the more innovative small firms. Optimising use of external sources is dependent upon participating in networks to identify market needs, leverage knowledge inputs from suppliers or absorb knowledge from other organisations, such as the research output from local universities.

Rammer et al did determine that some small firms involved in complex innovation do enter into R&D co-operation, alliances and joint ventures to access knowledge available from other organisations. The researchers concluded that continuous involvement in R&D is the main driver of innovation success in small firms. In contrast occasional involvement in R&D to merely solve a specific technological issue usually means the firm will not be successful as a sector innovator.

Recognition of the role of networks in providing access to new knowledge emerged in the 1980s (Hoang and Antoncic, 2003). Networks permit organisations to work together exchanging ideas, knowledge and technology (Heracleous and Murray, 2001), and can be a critical factor in survival and sustaining growth among small firms (Lechner and Dowling, 2000). Baum et al (2000) proposed that the purpose of vertical networks is to gain access to complementary resources; whereas the purpose behind horizontal networks is usually to achieve business growth.

Love and Roper (1999) determined that networking intensity had a positive influence on the level of innovations within UK manufacturing firms. MacPherson (1997), in a study of small US scientific instrument companies, concluded that internal R&D, when linked to external sources via networks, resulted in a higher level of innovation. Rogers (2004) suggested prior experience of success is likely to be reflected in a greater commitment to invest in network-based innovation. The degree of embededness within an innovation network is positively related to the inter-organisational exchange of knowledge and information (Fuentes et al, 2010). Hauser et al (2007) determined that the shape, size and purpose of business networks will undergo change over time. They proposed that networks exhibit a four phase development process; namely phase one, start-up; phase two, business formalisation; phase three, acquiring assets; phase four, innovation.

Camison and Villar-Lopez (2010) suggested participation in networks is not just limited to firms interested in innovation. Nevertheless Wincent and Westerberg (2005) found that small firms participating in networks do exhibit a higher degree of entrepreneurial behaviour. Only very limited published information is available however on whether network participation is beneficial in terms of sustaining performance during an economic downturn. Littunen and Virtanen (2009) did suggest that survival and growth during the current downturn were higher among Finish companies involved in business networks. Morfessis and Malachuk (2011) proposed that engaging in networking in an economic downturn does offer the benefit of being able to access additional scarce resources, thereby improving business survival rates.

Funding innovation

Case Aims: To illustrate the various ways small firms may be able to fund innovation projects.

For small firms engaged in high-technology product development, probably the most difficult task is raising the funding required to cover all phases of the project from idea identification through to market launch. The scale of funding will be increased in sectors such as medical devices where the developer will be required to engage in clinical trials to gain government approval for the product.

Various sources of funds can give possible support for technological innovation. One of the commonest is the founder of a company providing the funds, assisted in many cases by informal investors who are prepared to make capital available in return for a share of equity. Generally informal investors are acquainted with the company. Another source of funds may be government grants. Equity funding may be available from venture capitalists. The other source of funds can come from taking the company public. This usually cannot occur until after the new product has been validated as having market potential and market launch demands a significant injection of additional capital.

Rapoport (1990) examined a number of case examples to gain further understanding of small firm innovation funding in the US medical devices market. One company was initially funded by a group of physicians who were colleagues of the physician-inventor. Commercialisation was assisted by a research grant from the physician's hospital followed by further grants from the US government.

Another company came into existence following the initial product development undertaken by a polymer chemist working alone in his home workshop. He approached the president of a small healthcare consultancy business who personally funded the next phase of the development. The founder of a third company, with extensive experience of the equity markets, raised funds from a public offering very early into the development stage. In addition manufacturing development work was funded by an engineering company in return for being granted the rights to act as the contract manufacturer upon product launch. A fourth company obtained the majority of funds by selling the product's marketing rights to a major drugs company. Half the funding was made available immediately and the balance upon the new product receiving Federal government approval.

From his examination of these various cases studies, Rapoport concluded that in the medical devices market the commonest source of funds are those made available by the business founder; with cash flow from other products or business activities providing the next most common source of funds. Contributions from other organisations involved in the innovation process occur most frequently when a larger firm enters into a marketing agreement with the small firm. Another source of funds for some companies is to sell equity, most usually via a private placement with informal investors. Where larger sums are raised from a public securities sale, this is typically preceded by funds from private sources outside the established capital markets. Unlike the computer software industry small firms in the medical devices market rarely use venture capitalists as a source of funds. This would appear to be because the limited financial background of many company founders leads to problems during negotiations with venture capitalists. Founders tend to conclude that venture capitalists want an excessively large share in the business and this causes the entrepreneur to look elsewhere for money.

Radical innovation

For a small firm seeking to sustain long-term existence in a period of austerity the most effective strategy is probably to engage in radical innovation by developing a new product or process that uses significantly different technology (Chandy and Tellis, 1998). Radical innovation may lead to outcomes such as (i) market expansion, (ii) cannibalisation and/or (iii) destabilisation (Christensen, 1997). In relation to the total market expansion, radical innovations are those which offer a high potential for increased revenue. There is the key difference between the expected outcomes of radical versus incremental innovation. The latter tends to redistribute market shares within a sector. Radical innovation may deliver greater benefits to the founding company than were previously available and also substantially increase the size of the market. From the perspective of incumbent competitors that already have products within a category, radical product innovations may result in a higher probability of market destabilisation.

Assuming a small firm has the resources and capabilities to successfully implement radical innovation, the issue arises of the scale of competitive response, especially in relation to larger firms with greater resources. Although not specifically aimed at providing further understanding of this issue in the context of small firms, Aboulnasr et al (2008) used a study of the US pharmaceutical industry to review how firms respond to the advent of radical innovation. They proposed that the scale of response is likely to be greater when potential competitors observe firms introducing a radical innovation where the new product has a high potential to increase total market size. Large firms tend to require greater market potential than small firms. This is because smaller market opportunities tend to be of little interest to large firms in terms of the revenue benefits relative to the risks involved in engaging in radical innovation. As a consequence radical product innovation by a large firm is more likely to imply a much greater expectation of market expansion than similar activity by a small firm. Another influencing variable is degree of market dependence. Aboulnasr et al suggested where a firm is highly dependent upon an existing market and existing products are a major source of revenue, radical innovation is likely to be avoided due to concerns over product cannibalisation.

Aboulnasr et al's research of the pharmaceutical industry did reveal that, the greater the size of the firm introducing a radical product innovation, the higher is the probability that competitors will respond either by up-weighting promotional efforts for their existing products or seeking to replicate the radical innovation. The implications of this conclusion are that small firms engaged in radical innovation will initially be ignored during the early phases of the product life cycle (PLC), but should the innovation offer indications of major potential as the product moves into growth phase on the PLC then large firms will become a major source of competitive response. The researchers concluded that where large firm competitors are heavily dependent upon a specific market to generate the majority of revenue, this is also likely to result in a more aggressive response when confronted by radical innovation. The implication for the smaller firm is that where their success threatens a market sector which provides a major source of a large firm's revenues, the small firm should expect the large firm to retaliate very strongly to what will be perceived as a major threat to future sales.

The lower level of interest in small market segments often means these are ignored as sources of opportunity by large firms. As a consequence small firms can expect to operate in smaller markets without too many concerns about larger firms seeking to steal sales from them. Aboulnasr et al proposed that this situation may change in those cases where a large firm utilises radical innovation to enter what is currently perceived by the industry as a small market segment. Under these circumstances small firms need to be worried. This is because the arrival of a new large firm means the market segment is now perceived as potentially important and the large firm can be expected to be aggressive in seeking to acquire market share. The other problem is that, as other large firms observe these events, they may assume the market segment will become important in the future and decide to enter the segment as well, seeking to steal market share from existing small incumbents.

In large companies innovative ideas may originate from observing the behaviour of competition. In contrast the original idea within a small firm often is the result of reflective analysis in which the innovator draws upon personal experience and technical knowledge. The innovator's knowledge of industry in general and experience accumulated during the years in a particular industry usually increase the probability that the entrepreneur's idea will be extremely successful. The problem facing the small firm innovator is balancing resources between existing and new product development activities. In many cases strong revenue flows from existing products are critical because these provide the funding required to support new product development. An additional resource conflict can be the time needed by the innovator to manage the current business as well as being involved in new product development activities. In relation to this latter role, the innovator's time is usually required to identify relevant external actors and to build a viable network of subcontractors, co-operators or marketing agents (Oksanen and Rilla, 2009).

Small firm innovation

Case Aims: To illustrate some of the differences in radical innovation undertaken within a small versus a large organisation.

In terms of the core competences required to engage in effectively undertaking the innovation process Leonard-Barton (1992) proposed the following four dimensions:

(1) The knowledge and skills embedded in the employees.
(2) The knowledge embedded in technical systems (in relation to codified knowledge and procedures).
(3) The managerial systems that guide knowledge creation and control processes.
(4) The values and norms associated with undertaking innovation processes.

Small firms face a scarcity of financial and human resources accompanied by possible limitations over inadequate core competences. Mitigation to avoid these innovation-inhibiting factors may be achieved by involvement in networking to access external knowledge, resources and additional competences. To gain further insights into these issues, Pohl and Elmquist (2010) examined the development of a hybrid electric vehicle (HEV) at the relatively small Volvo car company and compared this with the much larger project undertaken by Toyota during the creation of the Prius.

Volvo developed the Desiree but decided against market launch. Nevertheless the development phase provides useful insights into how a smaller firm can overcome the scarcity of resources and competences. HEV is a radical technology capable of delivering the benefit of improved environmental responsibility through the use of electrical power to provide a driving experience comparable to a conventional petrol or diesel powered vehicle. Volvo had previously terminated their in-house hybrid development on cost grounds, but then Aisin, a Japanese firm, showed Volvo their new power-split hybrid transmission system. People involved in the Volvo HEV project obtained approval for a small-scale joint venture with Aisin. Other external knowledge was acquired by working with Varta to supply nickel metal hydride batteries and Bosch to supply an engine control module. The team was allowed a relatively high degree of freedom; thereby reducing the costs of administrative oversight. Development work was on a trial-and-error basis with suppliers complementing the technical competence within Volvo's small development team. In 1999 Ford, upon acquiring Volvo, were impressed by the HEV prototype and wanted the knowledge and technology to be transferred to their plant in Michigan. Subsequently the Volvo technology provided the basis for Ford's first HEV, the Escape.

A significant difference between the Toyota and Volvo projects was that the latter relied on external networking, whereas Toyota developed all core HEV technologies in-house. Toyota felt this reduced reliance upon outside suppliers and protected proprietary knowledge. The approach involved higher costs and the risk that an internal focus might cause the company to ignore external, possibly superior solutions to technical problems. Volvo decided to use existing vehicle chassis and conventional components as much as possible. This decision reflected a desire to keep overall development costs to a minimum. In contrast, all aspects of the Toyota's Prius design were entirely new.

Toyota's senior management ensured the project was given high status and priority. The team was provided with an extensive budget accompanied by access to resources and rapid decision making. The Volvo project was run by a small dedicated team with a very limited budget, who needed to persuade the rest of the company that hybrid vehicles was a strategic opportunity. To gain access to resources and support of departments within the company, the Volvo team had to exhibit a much higher level of entrepreneurial competence than their counterparts at Toyota.

Business process innovation

Kim and Mauborgne (2005) advocated that a strategy based upon the introduction of a new business model of the type utilised to support a blue ocean philosophy is superior to other forms of innovation. Their rationale was that this strategic approach permits firms to access new forms of customer demand; thereby outperforming competitors who remain focused upon sustaining their existing business model in existing markets. Markides and Geroski (2005) were also supportive of the idea that business model innovations which challenge conventions will lead to superior performance. Szulanski and Jensen (2008) presented a slightly different perspective. They proposed that superior profits greater than those available from initial innovation can be achieved by the innovator firm learning how to subsequently refine and develop a new business model. This latter action permits the firm to replicate the model in new geographical locations and to repeat the process when developing the next generation of products. McDonald's and Starbucks are examples, suggested by Szulanski and Jensen, of companies which have implemented geographic replication of a proven new model when expanding overseas from their original domestic American market. A proposed example of next generation replication is exemplified by Intel's success in the development and launch of their next generation microchips.

Aspara et al (2010) posited that business model innovation, accompanied by low emphasis on replication, generates higher profitability than not engaging in this activity. They concluded however that because small firms have limited resources these organisations are less able to enhance financial performance through replication of business model innovation in new markets or for next generation products. These authors suggested that in an increasingly unstable world more small firms should attempt to place greater emphasis on business model innovation replication. In their view, the flexibility and proactive orientation of the more entrepreneurial small firm should permit the organisation to outperform competitors who remain fixated on continuing to rely on their long-established, existing business models.

Aspara et al researched the performance of different size firms in Northern Europe. Data from the study supported the perspective that small firms with a high strategic emphasis on business model innovation but low on replication have higher average profitable growth than small firms with low strategic emphases on business model innovation and replication. The average profitable growth of small firms with high strategic emphases on both business model innovation and replication did not significantly differ from the average profitable growth of small firms that emphasised business model innovation but gave low emphasis to replication.

Radically new products have the potential to obsolete existing products and this may act as an innovation disincentive among large incumbent firms. This outcome is known as the 'incumbent's curse'. Non-incumbents may enjoy an advantage in terms of being first movers in the exploitation of radical product innovations. Large firms can be prone to the forces of 'bureaucratic inertia', because as the number of employees increase this may cause the organisation to introduce additional layers of management and more formalised policies and procedures (Tornatzky and Fleischer, 1990). As a consequence small firms may possibly be in a better position to introduce radical product innovations than large firm incumbents. Examples of this phenomenon are provided by the early years of firms such as Amazon in online retailing, YouTube in online video streaming and Facebook in social networking. Chandy and Tellis (2000) undertook a large-scale study of the consumer durables and office equipment market in the USA. They concluded that although in certain sectors small firms still lead in the introduction of radical innovation, in recent years this pattern has begun to change. They concluded that this outcome reflects the greater willingness of large US firm incumbents to engage in radical innovation and are also now more accepting of obsoleting their prior generation products. One possible reason for this behaviour shift is that many large firms now operate in the rapidly changing world of high technology and have come to understand that retention of market leadership will require product cannibalisation to stay ahead of the competition (Chandy and Tellis, 1998). Another possible factor of influence is that some large Western companies have now moved towards decentralisation and the creation of smaller, autonomous organisational units. This structural change has thereby enabled these firms to more rapidly exploit new technology whilst maintaining leadership over current technology until the new radical innovation is available for market launch. Another advantage enjoyed by this type of large firm is their breadth of technological capability linked to large R&D operations. This results in these firms now being more likely to become aware first of new scientific breakthroughs accompanied by the added advantage of being in a stronger financial position to pursue radical product innovation than their smaller counterparts.

References

Aboulnasr, K., Narasimhan, O., Blair, E. and Rajesh Chandy, R. (2008), Competitive response to radical product innovations, *Journal of Marketing*, Vol. 72, pp. 94–110.

Aspara, J., Hietanen, J. and Tikkanen, H. (2010), Business model innovation vs replication: financial performance implications of strategic emphases, *Journal of Strategic Marketing*, Vol. 18, No. 1, pp. 39–56.

Atuahene-Gima, K. (1996), Market orientation and innovation, *Journal of Business Research*, Vol. 35, No. 2, pp. 93–103.

Baum, J., Calabrese, T. and Silverman, B. (2000), Don't go it alone: alliance network composition and start-ups' performance in Canadian biotechnology, *Strategic Management Journal*, Vol. 21, No. 3, pp. 267–294.

Camison, C. and Villar-Lopez, A. (2010), Knowledge management by external links effect of SMEs' international experience on foreign intensity and economic performance, *Journal of Small Business Management*, Vol. 48, No. 2, pp. 116–142.

Chandy, R.K. and Tellis, G.J. (1998), Organizing for radical product innovation: the overlooked role of willingness to cannibalize, *Journal of Marketing Research*, Vol. 35, pp. 474–487.

Chandy, R.K. and Tellis, G.J. (2000), The incumbent's curse? Incumbency, size, and radical product innovation, *Journal of Marketing*, Vol. 64 , pp. 1–17.

Chaston, I. (2010), *Entrepreneurship and Small Firms*, Sage, London.
Christensen, C.M. (1997), *The Innovator's Dilemma*, Harper Business, New York.
Fuentes, M.M., Arroyo, M.R., Bojica, A.M. and Pérez, V.F. (2010), Prior knowledge and social networks in the exploitation of entrepreneurial opportunities, *International Entrepreneurship and Management Journal*, Vol. 6, No. 4, pp. 481–501.
Hauser, C., Tappeiner, G. and Walden, J. (2007), Learning region: the impact of social capital and weak ties on innovation, *Regional Studies*, Vol. 41, No. 1, pp. 75–88.
Heracleous, L. and Murray, J.A. (2001), Networks, interlocking directors and strategy: towards a theoretical framework, *Asia Pacific Journal of Management*, Vol. 18, pp. 137–160.
Hoang, H. and Antoncic, B. (2003), Network-based research in entrepreneurship: a critical review, *Journal of Business Venturing*, Vol. 18, No. 2, pp. 165–187.
Jeffries, A. (2011), 4 ways to bounce back, *Black Enterprise*, Vol. 41, No. 11, pp. 42–44.
Kim, W.C. and Mauborgne, R. (2005), *Blue Ocean Strategy: How to Create Uncontested Market Space and Make the Competition Irrelevant*, Harvard Business School Press, Boston, MA.
Kohli, A.K. and Jaworski, B.J. (1990), Market orientation: the construct, research proposition, and managerial implications, *Journal of Marketing*, Vol. 54, No. 2, pp. 1–18.
Lechner, C. and Dowling, M. (2000), The evolution of industrial districts and regional networks: the case of the biotechnology region Munich/Martinsried, *Journal of Management and Governance*, Vol. 19, No. 3, pp. 309–338.
Leonard-Barton, D. (1992), Core capabilities and core rigidities: a paradox in managing new product development, *Strategic Management Journal*, Vol. 13, pp. 111–125.
Littunen, H. and Virtanen, M. (2009), Differentiating factors of venture growth: from statics to dynamics, *International Journal of Entrepreneurial Behaviour & Research*, Vol. 15, No. 6, pp. 535–547.
Love, J. and Roper, S. (1999), The determinants of innovation: R&D, technology transfer and networking effects, *Review of Industrial Organisation*, Vol. 15, pp. 43–64.
MacPherson, A.D. (1997), A comparison of within-firm and external sources of product innovation, *Growth and Change*, Vol. 28, pp. 289–308.
Markides, C. and Geroski, P.A. (2005), *Fast Second: How Smart Companies Bypass Radical Innovation to Enter and Dominate New Markets*, Jossey-Bass, San Francisco.
Morfessis, I. and Malachuk, D. (2011), Economic development in the post crisis era, *Economic Development Journal*, Vol. 10, No. 3, pp.14–22.
Narver, J.C., and Slater, S.F. (1990), The effect of a market orientation on business profitability, *Journal of Marketing*, Vol. 54, No. 4, pp. 20–35.
Nash, B.J. and Zeuli, K. (2011), Small business lending during the recession, *Richmond Federal Economic Brief*, Virginia, Vol. 11, No. 2, pp. 1–6.
Oksanen, J. and Rilla, G. (2009), Innovation and entrepreneurship: new innovations as a source for competitiveness in Finnish firms, *International Journal of Entrepreneurship*, Vol. 13, pp. 36–45.
Palmer, J. and Griswold, M. (2011), Product and service innovation within small firms: an exploratory case analysis of firms in the restaurant industry, *International Journal of Business and Social Science*, Vol. 2, No. 13, pp. 34–41.
Petrou, A. and Daskalopoulos, I. (2009), Innovation and small firms' growth prospects: relational proximity and knowledge dynamics, *Planning Studies*, Vol. 17, No. 11, pp. 1592–1602.

Pohl, H. and Elmquist, M. (2010), Radical innovation in a small firm: a hybrid electric vehicle development project at Volvo Cars, *R&D Management*, Vol. 40, No. 4, pp. 372–384.

Rammer, C., Czarnitzki, D. and Spielkamp, K. (2009), Innovation success of non-R&D-performers: substituting technology by management in SMEs, *Small Business Economics*, Vol. 33, pp. 35–58.

Rapoport, J. (1990), Financing of technological innovation by small firms: case studies in the medical devices industry, *Small Business Economics*, Vol. 2, pp. 59–71.

Rogers, M. (2004), Networks, firm size and innovation, *Small Business Economics*, Vol. 22, pp. 141–153.

Saridakis, G., Mole, K. and Storey, D. (2008), New small firm survival in England, *Empirica*, Vol. 35, No. 1, pp. 25–39.

Storey, D. (2002), Methods of evaluating the impact of public sector policies to support small business, *International Journal of Entrepreneurship Education*, Vol. 1, No. 2, pp. 180–202.

Szulanski, G. and Jensen, R.J. (2008), Growing through copying: The negative consequences of innovation on franchise network growth, *Research Policy*, Vol. 37, pp. 1732–1741.

Tornatzky, L.G. and Fleischer, M. (1990), *The Process of Technological Innovation*, Lexington Books, Lexington, MA.

Verhees, F.J.H. and Meulenberg, M.T.M. (2004), Market orientation, innovativeness, product innovation, and performance in small firms, *Journal of Small Business Management*, Vol. 42, No. 2, pp. 134–154.

Wincent, J. and Westerberg, A. (2005), Personal traits of CEOs, inter-firm networking and entrepreneurship in their firms, *Journal of Developmental Entrepreneurship*, Vol. 10, No. 3, pp. 271–284.

14
The Public Sector

Creating the problem

The Great Depression of the 1930s led to mass unemployment and declining standards of living across much of the Western world. This influenced political thinking, and the end of World War II saw the creation of the large welfare state which now absorbs the vast majority of public sector spending within the Western democracies (Lindbeck, 1995). Even as early as the 1980s, Western politicians were becoming aware that the combined influences of population ageing and increasing costs of healthcare provision would eventually lead to the situation where sustaining the public sector would cease to be a financially viable option. Few politicians articulated their concerns over this 'ticking time bomb', presumably because of the risk of losing the support of their electorate.

In the 1980s politicians were usually more concerned about inflation, which they perceived as the most dangerous factor impacting the economic survival of nations. As a consequence most governments accepted the need to assign Central Banks the role of using monetary policy to combat inflation (Schettkat, 2001). Unfortunately, despite warning from Central Bankers about the concurrent need to control public sector spending, most industrialised countries continued to run persistent deficits leading to rapidly rising debt-to-GDP (gross domestic product) ratios. This deterioration in fiscal balance sheets was mainly due to politicians continuing to approve expansion of public sector expenditure in areas such as unemployment benefits, pensions and healthcare (Weiner, 1995).

By the early 1990s the banking industry, especially in the USA and the UK, had successfully lobbied politicians to remove regulations, such as the Glass-Steagall Act of 1933, designed to avoid a repetition of banking collapses of the type which occurred during the Great Depression. Politicians' confidence in avoiding another banking crisis was based upon an assumption that Central Banks and other regulatory agencies would provide early warnings of potential problems within the financial services sector (Kaufman and Wallinson, 2001). As part of their anti-inflation and economic stability policies, Central Bankers kept interest rates very low. What governments did not seem to understand, or preferred to ignore, is when the world is awash with money due to uncontrolled lending supporting ever increasing consumer spending, the inevitable outcome is economic instability (Nesvetailova, 2005).

The sub-prime mortgage problem in the USA triggered a global banking crisis. This subsequently forced some governments to take over banks and inject money into their economies to avert an economic disaster (Hoenig, 2008). As the scale of financial crisis worsened, it became apparent that some countries such as Greece had been growing their economy by increased public sector spending funded through borrowing low interest short-term loans (Bauer et al, 2003). When liquidity began to rise as the banks reduced involvement in inter-bank lending, some countries could no longer obtain replacement funding at an affordable cost when these loans became due. America, the UK and some European governments accepted the need to reduce the size of their respective public sector deficits, despite the adverse impact on the provision of welfare state services. In the case of countries such as Ireland and Greece, a reduction in public sector spending was a mandatory requirement of the IMF (International Monetary Fund) and the ECB (European Central Bank) before these institutions were willing to make funds available to support the rebuilding of these nations' shattered economies. The net effect of all of these events is that public sector organisations (PSOs) in virtually every Western democracy can expect to face a lengthy period of financial constraint as their respective governments struggle to reduce public sector deficits. How long this period of public sector austerity will last is difficult to predict. It would seem likely that little improvement in the scale of the deficits can be expected for at least another five years and that, in some European countries, any significant deficit reduction will not be achieved until well past 2020. This outcome will inevitably prolong the length of the new austerity in the affected countries. Because, in areas of the world such as the Mediterranean, reduction in public sector spending will adversely impact inhabitants' spending levels, this in turn will have a negative multiplier effect on world trade, extending the period of austerity even in nations not engaged in public debt restructuring.

Cutback response

The response of the majority of PSOs in those Western democracies struggling to reduce financial deficits has been to reduce the provision of welfare services and make staff redundant. Many PSOs have community and service-orientated goals that appeal to the communities they serve (Wright, 2007). This typically leads to enduring support among the general public for PSOs engaged in delivering welfare state services; but such support is likely to decline when the general public is required to pay more taxes or personally contribute towards the cost of service provision.

Making staff redundant threatens PSOs' social contract with their employees, which poses a direct challenge to sustaining employee motivation. A decline in employees' trust in management will follow and those employees with vital skills in specialty occupations are usually the first to seek alternative employment. Staff members who remain are often faced with rising levels of stress as they seek to sustain their commitment to delivering services with fewer resources. The scale of the human resource management (HRM) problems created by such cutbacks does raise major questions over the advisability of adopting a reductionist approach to handling budget crises. This is because the approach runs the risk of either solving the wrong problems or, over time, making the current problems even worse (Pandey, 2010).

Reform required—but will it happen?

Case Aims: To illustrate how one area of the public sector has done little to reduce the cost of service provision.

Most organisations facing costs rising faster than income would seek to become more efficient and avoid incremental expenditure, which cannot lead to improvements in productivity. This rationale however would not appear to be the case in relation to the UK higher education sector. Over recent years expenditure on administrative functions within many UK universities has increased, whereas funding of teaching has remained virtually unchanged. One of the largest increases in administrative expenditure has been on salaries for senior staff. The response to any criticism over this trend is that universities are complex organisations and hence require the services of senior management capable of providing outstanding leadership. Even if one accepts the validity of the argument that these organisations are difficult to manage, there still remains the question of whether the standard of required leadership necessitates paying, in some cases, a salary twice that of the UK prime minster.

The response by the sector to questions over increases in expenditure is that there now exist many more regulations, such as protecting the rights of students, a requirement to meet government key performance indicators (KPIs) and ever greater demands in relation to sustaining the quality of the learning experience. When one examines how such administrative tasks are undertaken it is apparent however, that little effort has been made to find ways of improving the management of internal processes. Many institutions have a diversity of different computer systems, often containing software legacy problems requiring manual transfer of data between systems. Few institutions have adopted the widely validated automated paperless data processing technologies which are now common in the private sector for optimising work flows. Furthermore these institutions are increasingly reliant upon the use of the Internet to support student learning. Yet most universities still insist on managing their own IT systems, reject the cost effectiveness of cloud computing and have website downtime problems the scale of which would bankrupt any commercial organisation involved in the utilisation of online data transfer.

Few question that the UK is facing the worst economic crisis since the 1930s Great Depression. Most people accept that the scale and speed of the economic recovery will be influenced by the number of people with the appropriate higher level competences and skills entering the future work force. However despite the fact that student loans are available to cover the costs of higher tuition fees, more school leavers and many of their parents are questioning the benefits of continuing education past the age of 18. As a consequence more school leavers are deciding they can no longer afford to attend university. Over time this trend will be detrimental to the supply of higher knowledge skills individuals entering the UK workforce.

The way the sector is currently managed raises the question over whether the UK is receiving value for money from the nation's universities. This question is not unique to the UK. In recent years similar concerns have been raised about how the sector is managed in the USA. Some of the sector's greatest criticisms have come from leaders in the country's high-technology industries such as the late Steve Jobs and Bill Gates. These individuals cannot understand why the sector has failed to exploit the opportunities for making degrees more affordable and the increase in quality of programme provision through the exploitation of computers and online technologies. This critical perspective is not just isolated to individuals from the private sector. The ex-president of Harvard University, Professor Larry James, noted that the only organisations which have failed to respond to major environmental changes over the last 100 years are the universities.

Entrepreneurship

Although reducing services and making staff redundant is the standard public sector response when faced with budget cuts, the question arises of whether this is the only option that has been considered by these organisations. Available alternative options include examining how the organisation might become more efficient, identifying alternative revenue sources and staff resource being re-deployed.

The new public management (NPM) concept, which emerged in the late 1980s, was designed to improve effectiveness and efficiency within the public sector. Some academics proposed that NPM strategies could be made more effective by stimulating entrepreneurial orientation to improve services and upgrade organisational productivity (Morris and Jones, 1999). Brooker (2005) described the public sector entrepreneur as an individual prepared to go beyond implementing a reactive response to changing circumstances and enact a pioneering, imaginative approach to resolving operational problems and resource shortages.

Unfortunately attempts to introduce innovative solutions in the public sector are often impaired by the parochialism and differing vested interests of staff inside the organisation. As a consequence attempts to transfer entrepreneurial processes from the private sector without including modifications to reflect the different cultures within the two sectors are probably doomed to fail (Moore, 1995). Morris and Jones (1999) proposed that in considering entrepreneurship strategies recognition must be given to the fact that PSOs (i) are funded by taxation, (ii) have less exposure to genuine market forces, (iii) are providers of services often not available from any other source, (iv) are subject to public scrutiny and (v) are faced with risk–reward trade-offs that strongly favour mistake avoidance. The other problem is the concern among politicians and senior civil servants that granting PSOs greater freedom and autonomy may permit the emergence of 'rogue' entrepreneurs, whose lack of integrity could result in outcomes such as the misuse of public funds, aggressive coercive pressures applied to dissenters and a failure to understand that radical change cannot succeed by ignoring sector traditions and conventions (DeLeon and Denhard, 2000).

Currie et al (2008) interviewed CEOs from different areas of the UK public sector and concluded that the primary focus of innovation was on seeking better use of limited resources. Thus entrepreneurial activity is often driven by government financial agendas in relation to delivering specified policies and meeting required outcomes defined by KPIs. The CEOs accepted the idea of risk taking but were keenly aware that the public sector is a very unforgiving environment in which to make mistakes. As a consequence employees tend to block rather than support proposed change unless they perceive there are ways of ensuring that any subsequent blame can be laid at the door of others within the organisation. Another problem which can compound an aversion to innovation is that many employees currently operate in a climate of uncertainty and job insecurity. This can cause paralysis among staff members because they are more concerned about avoiding redundancy than seeking ways of enhancing fulfilment of assigned responsibilities.

Public sector innovation may occur for a variety of reasons. These include politically-led responses to crises, organisational turnarounds engineered by new leadership or bottom-up innovations initiated by front-line staff or middle managers (Borins, 2002). Social, political/legal, economic and technological environments often act as external sources of change. Their influence depends on the nature of the organisation's role, the scale of environmental change and the power of external stakeholders. Changes in the beliefs, values, attitudes and opinions of people can serve as a source of social changes. Governmental changes or new laws are political/legal changes that may have impact in relation to the scale and terms under which funding is made available to a PSO.

Chaston (2001) suggested there are two types of innovation; namely anticipatory and reactive. Anticipatory decision makers can be considered as pioneers injecting a new entrepreneurial orientation into a PSO's operations. Most managers however exhibit a reactive response only implementing change in the face of intense external or internal pressures. Reger et al (1994) proposed the term 'fundamental change' to describe actions that alter the very character of the organisation and contrasted this with 'incremental change', which only involves minor revisions in processes, structure or strategy. The whole organisation is affected by fundamental change; whereas incremental change usually involves minor adjustments to current activities and typically occurs while external and internal environments are relatively stable.

In attempting to implement innovative ideas in the public sector entrepreneurs can expect to encounter one or more of the following barriers to potential change:

(1) Knowledge barriers resulting from a lack of information regarding a task or a change in plan.
(2) Skill barriers due to the lack of competences necessary to implement proposed change.
(3) Commitment barriers involving a refusal to accept the change process because of contrary objectives or lack of motivation.
(4) Institutional barriers caused by values, regulations and culture within an organisation influencing employee behaviour.
(5) System barriers in the form of inadequate resources to implement change projects.

Promoting innovation

Case Aims: To illustrate the role of promoters in the management of change involving innovation.

Husig and Mann (2010) proposed that a critical aspect of entrepreneurial change, especially in the context of a PSO, is the presence of one or more 'promoters'. These are individuals who actively assist the innovation process. One type of promoter are individuals who have the hierarchical power to overcome barriers by exerting their position power, punishing opponents or protecting the innovators. The knowledge promoters are individuals who utilise their expertise and reputation to convince opponents to accept the benefits of change. This latter type of individual uses relationships with key personalities to build networks and link together promoters and employees from different hierarchy levels or units (Lines, 2007).

Husig and Mann researched the change process in a German university where growing sectoral competition for students provided a strong economic incentive to implement programmes to improve reputation and educational quality. Another external driver was technological developments that enable the offering of e-learning, preferably ahead of the competition. Drivers were examined in the context of the university's internal promoter for change. His background permitted fulfilment of the role of being the knowledge promoter. A major external barrier was the Bavarian Ministry of Science, which had to be convinced of the need for change in order to pass legislation to permit the university to make fundamental revisions in operational processes. Inside the university an internal barrier was the Senate, which initially resisted the idea of implementing proposed changes. This outcome reflected the Senate's lack of understanding of why change was necessary.

Once the new approach to educational provision was approved by the Senate, system barriers associated with inadequate resources caused delays in programme implementation. The lead promoter used his knowledge to gain understanding among staff of the benefits of innovation to overcome these barriers. Another critical promoter was the university rector who used his network to establish links between the lead professor and the Bavarian minister of science as a path for acquiring additional resources from the government.

Technological innovation

Although given less emphasis by the strategic planner in many PSOs, new technology can have a major impact on changing the nature of services provision or enhancing the efficiency, effectiveness and economics of service provision. Given the primary activity of many PSOs is the acquisition, storage, analysis and implementation of actions involving large quantities of information, the most obvious entrepreneurial opportunities are those related to the exploitation of IT (Herps et al, 2003).

Entrepreneurial PSOs will be more aware of new opportunities offered by technology and incorporate these into the development of future strategies. More conservative organisations may place less importance on the role of technology in the development of new services or upgrading internal processes. Awareness of the importance of technological change in PSOs tends to be focused on investment actions to increase productivity and replace high-cost labour inputs with automated, machine-based processes (Chaston, 2011). Although this strategic orientation can significantly improve operational performance, there is the risk that excessive emphasis on process change will mean opportunities to make available new services may be ignored.

PSOs have a poor track record in correctly identifying the most appropriate new technology to incorporate into their operations. Frequently the development phase results in cost overruns or the technology not delivering expected cost–benefit outcomes specified when contracts with suppliers were originally agreed. In some cases the scale of implementation problems proves to be so large that eventually the project is terminated before completion with a resultant major loss of public sector monies.

As noted by Goldfinch (2006), some of the most expensive failures associated with the introduction of new technology into the public sector, leading to the waste of vast sums of money, have been IT projects. In his view, in the USA the majority of public sector projects to significantly upgrade information management systems have been unsuccessful. Furthermore the larger the project, the greater is the probability that the project will fail to deliver the cost–benefits which were claimed at the project outset. Such failures are not restricted to the USA. In the UK the Wessex Health Authority's Regional Information Systems Plan was cancelled after spending what is estimated to be in excess of £40 million on a non-functional system. The New Zealand Police abandoned an information systems development in 1999 at a cost of more than $NZ100 million.

There are various reasons why IT projects fail. In some cases the system never performs at standards specified in the original contract. Alternatively, even when the system appears to be functioning correctly, the project may not generate productivity gains or deliver forecasted benefits. Another failing can be the system never being brought on-stream due to internal user resistance reflecting factors such as managerial recalcitrance, lack of staff training or the inability of staff to cope with the complexity of the new system.

Goldfinch believes a key factor influencing poor project outcomes is knowledge asymmetry. This is because the public sector project manager usually has much less understanding of the technology than the suppliers' representatives. These latter individuals are thus in a position to possibly conceal from the client that the project is likely to face fundamental development problems. Furthermore it is often the case that the public sector project leader is reluctant to be the bearer of bad news and hence may downplay emerging problems when reporting to senior management or an independent public sector monitoring agency.

One of Goldfinch's suggestions, where the decision over a new IT investment is being considered, is that the first question to be asked is how this project can be undertaken with the least disruption, cost, risk and uncertainty. The most dangerous course of action is to invest in high-risk, highly-ambitious projects involving leading-edge technology or lengthy development time frames. This is because the probability of failure in these latter situations is very much higher. In Goldfinch's opinion, the most sensible solution is to examine what is currently working in the private sector and to buy an off-the-shelf system which has already been demonstrated to work properly. In those cases where such a system is not available, then possibly the best option is to accept that technology is not yet widely available that can produce the sought IT solution.

Some PSOs have successfully exploited off-the-shelf technology and commercially available database software to enter the 'digital age' of service provision. Costs have been reduced, speed of response increased, new websites established and electronic procurement systems created to support a transition from a paper-based to an electronic information management world. Dunleavy et al (2005) concluded that the benefits digitalisation has brought to the public sector include capture of economies of scale and scope, radically reduced operating costs and upgraded back-office administration functions. Furthermore as online systems have assisted in the integration of functions this has assisted PSOs to become 'one-stop-shop' service providers.

Digitalisation can be utilised to rapidly implement significant public sector innovation at relatively low cost. The transport authorities in London decided to install automated charging technology in underground rail stations and buses using a smart card (the 'Oystercard'). Users add credit to their card and then pay for any journey by swiping the card past an automatic reader. Initially 350,000 existing holders of paper-based season tickets switched to the electronic card. As awareness of the benefits of the Oystercard became apparent, card usage grew to over 2 million people in less than four years. This expansion led to large cost savings in ticketing staff, major reductions in peak-hour queuing times and increased the use of mass transit by passengers for whom ticket acquisition was no longer a potential problem within the journey process.

Factors of influence

Case Aims: To illustrate the nature of factors that can influence the effective implementation of technology innovation projects.

Parnaa and von Tunzelmann (2007) investigated technology-based innovation project implementation across four different European countries. They concluded that the most important innovation goal is to respond to user needs; followed by improvements in service quality, taking the organisation online and improving service take-up. The least important innovation goal was improvements in the organisation's competitiveness or internal organisational behaviour.

Variation existed in different countries' innovation goals. In Estonia the most important goal was to reduce the time spent on service delivery, in Finland and the UK to go online and in Denmark to improve service transparency. Internal factors influencing successful project outcomes were personal leadership, individuals with appropriate technical skills, top management commitment and open-minded staff. Close co-operation with technology suppliers and good knowledge of the existing technologies were also rated as important. Hierarchical top-down power was rated by all countries as the least important factor for project success. In relation to different internal factors hampering innovation, the most important was the lack of knowledge about existing technologies, followed by poor co-operation with technology suppliers, weak top management commitment and a non-supportive organisational strategy.

Parnaa and von Tunzelmann concluded that, from a managerial perspective, the factors influencing the innovation process in PSOs differ somewhat from those in the private sector. Competitiveness and service cost are important innovation drivers in the private sector. Both of these factors were among the least important innovation goals in the public sector. In terms of differences between countries, the researchers concluded that the large size of the UK resulted in the existence of bigger and more complicated institutional set-ups and services. As a consequence the innovation process is riskier and more costly, resulting in the need for stronger top-management commitment and support, closer co-operation with technology suppliers and better understanding of the market demand for services.

Prevention innovation

The adage 'prevention is better than cure' has, in various forms, existed for hundreds of years. The primary purpose of the phrase is to suggest that taking action to prevent or avoid an illness may be tiresome and even possibly painful, but nevertheless this is usually a more preferable outcome than individuals having to subsequently endure a lengthy treatment for an actual medical condition. Although the philosophy primarily is utilised in relation to the provision of healthcare, equivalent concepts are also applicable in others areas of the public sector such as crime prevention, education and the provision of social services. In a period of prolonged public sector spending austerity, the issue arises of whether allocation of resources to prevention, initiated as a result of innovation, is more cost effective than the ongoing provision of services associated with resolving the actual situation faced by affected individuals.

An area of the public sector expenditure where successful prevention has the potential to greatly reduce total spending levels is crime. Currently tax payers are funding extremely high expenditure to support policing in order to identify perpetrators, operate a judicial system and run prisons. There are the added indirect costs of the adverse impact of crime upon victims and damage to the economic prosperity of a country. Over the years various initiatives have been implemented with the aim of reducing crime levels, such as increased policing levels, crime prevention partnerships between the police and local communities and technology such as CCTV for monitoring events in high crime-incidence areas. The general public have a strong desire to continue to observe a heavy police presence on their streets and police forces are highly motivated to catch criminals. Hence in an era of public sector cutbacks the issue over whether funding cuts should be made to crime prevention versus crime detection initiatives is expected to result in supporters for these two activities fighting to retain budgets for those areas of the police and justice system which they think is most effective.

The other problem in relation to crime prevention is what actions society or their political representatives are willing to accept in terms of possible deterrents and punishments. The greatest cause of crime in many nations is the use of illegal drugs. One way of reducing the crime rate for drug-related offences would be to legalise all illegal substances, making them available either as marketed products similar to cigarettes and alcohol or by the establishment of medically monitored and controlled usage centres (Wolf, 2011). The concept of legalising drugs was presented in the UN's recent publication entitled the *Global Commission on Drug Policy* (Wolf, 2011). Among the signatories of this report were George Schulz, former US secretary of state; Paul Volcker, former chairman of the US Federal Reserve; Kofi Annan, former UN secretary general; Ernesto Zedillo, former president of Mexico; and Fernando Cardosa, former president of Brazil. All these respected and highly experienced individuals supported the view that the criminalisation of drugs has led to huge levels of corruption among public officials and the destruction of significant proportions of civilised society, especially in those Central American countries embroiled in the ongoing drugs wars between the state and the traffickers. As reflected by the adverse response in the media to the report, society's prevailing perspective is that proposals are complete heresy and implementation would lead to the complete meltdown of society. As a consequence politicians in Western nations hoping to retain the support of the electorate are unlikely to propose legalising drugs, even though this prevention strategy would have a very significant impact on reducing crime levels, as well as lowering expenditure on policing, the criminal justice system and the operation of prisons.

The issue of prevention versus financing the resolution of subsequent adverse outcomes is also apparent in public sector service provision activities such as education, child poverty and helping problem families. In all of these sectors, in most Western nations, many existing initiatives face funding cutbacks and a probable decrease in the number of new more innovative approaches coming on-stream in the foreseeable future. An additional obstacle to introducing new approaches is the frequent resistance of those engaged in delivering existing programmes to the idea of resources being transferred elsewhere either within their own departments or to another PSO. Nevertheless this situation should not be seen as a reason to reject adoption of an entrepreneurial orientation towards any form of prevention innovation. Given that many existing prevention projects are very labour intensive, possibly the area which offers the best cost–benefit outcomes is the use of new IT-based solutions for predicting crime probabilities, telemedicine for the remote monitoring of patients and the provision of customised computer-based software to assist children with learning difficulties.

Healthcare

Healthcare prevention activities can be divided into three categories: primary, secondary and tertiary. Primary prevention involves the reduction or control of causative factors (e.g. vaccinations to prevent diseases). Secondary prevention is associated with the earliest possible detection, which permits more successful treatment outcomes (e.g. cancer screening services). Tertiary prevention involves detection of secondary complications among people already being treated for a primary illness (e.g. screening for circulation and retinal problems among diabetics) (Teutsch, 1992).

The issue of the costs–benefits of prevention versus cure are rarely simple, because there is the requirement to weight the cost of prevention against the estimated cost of the future treatment. Additionally switching from cure to prevention without continuing to sustain expenditure on the provision of the cure component is rarely a feasible proposition. This is because the outcomes of prevention programmes often take many years to have real impact and during this time period cure provision expenditure will still remain high. A further problem in a world of scarce resources is how to fund a new prevention initiative. In those cases where the proposed source is to reduce expenditure on treatment activities there may be significant resistance to this solution from those currently engaged in the delivery of cure-based care activities.

Childhood immunisation offers an example of a proven cost-effective prevention programme. Certain types of neonatal screening are also cost effective because the costs of tests for certain congenital disorders save money relative to the much higher costs of treatment at a later date (Leutwyler, 1995). As far as expenditure on many of the other available screening programmes goes, the literature suggests these are unlikely to be cost effective when delivered as a mass screening programme to an entire population. This is because most illnesses have low incidence rates and hence screening everybody will only lead to the identification of the specific medical condition being tested among a very small number of people. Cost effectiveness can be improved however by more careful targeting of those to be screened, an outcome which occurs, for example, when one restricts breast screening to women aged 50 and over (Stanaland and Gelb, 1995).

Another contentious issue in relation to prevention-based cost-reduction healthcare strategies is that, by preventing a condition through immunisation or providing treatment in the early stage of an illness, people will then live longer. Although this is good news for the individual, the bad news is that as people live longer they are more likely to need treatment for a much more expensive medical condition, such as an organ transplant, or years of care because they have developed an age-related mental illness like dementia or Alzheimer's.

A major issue of concern to many healthcare professionals is the rising level of obesity in many developed nations, not just in adults, but increasingly amongst children. Factors of influence here include diet and sedentary lifestyles. The problem with obesity is that the condition can lead to other medical problems, such as diabetes and circulatory problems, later in life. This situation has resulted in increased emphasis on prevention policies, such as ensuring children only have access to healthier meals when at schools, more focus on promoting increased exercise and educating parents over the need for healthier eating both at home and when eating out. An identified obstacle in many of these initiatives however is that amongst the poorest of families declining incomes, which increase during a period of austerity, will usually result in rising consumption of high carbohydrate diets.

The obesity paradox

Case Aims: To illustrate the potential financial contradictions that can emerge in relation to healthcare strategies designed to prevent versus treat a medical condition.

In terms of actions to reduce obesity, whether these are undertaken in schools or by the providers of healthcare, such activities currently represent an area of rising incremental public sector costs. The issue arises as to whether this expenditure results in an overall reduction in healthcare delivery costs. Rappange et al (2009) examined differences in lifetime healthcare costs between the obese and 'healthy-living' cohorts. Their assumption was that a prevention strategy that converts obese people into individuals who embrace healthier lifestyles is a costless proposition. From their analysis they concluded that, in the first 50 years, converting obese people into healthier people will save costs in all areas of healthcare provision by lowering disease incidence rates. The largest savings occur in the costs of drugs and expenditure on hospital care. When people reach the age of 50, however a different picture emerges. This is because preventing obesity results in the incurrence of additional costs in life-years gained in all healthcare segments, most especially in the long-term care sector. As a consequence, over the longer term, preventing obesity may increase rather than decrease a nation's total healthcare costs.

In terms of whether such findings should alter healthcare strategies over preventing obesity, two issues arise. Firstly by the time young people reach the age of 50 advances in other areas of medicine may have reduced care costs for the elderly. The other, possibly more important, issue is the societal costs and benefits which need to be considered in any cost–benefit analysis. For example, reduced obesity may result in less obesity-related morbidity and higher life expectancies, which in turn might improve the productivity of individuals and their contribution to society through employment and being more effective parents. Furthermore as the proportion of elderly in developed country populations continues to increase, the emergence of a greater proportion of much healthier people of pensionable age might lead to a reduced strain on the delivery of formal care in residential homes for the elderly and a stronger network of adults able to care for their ageing parents.

Given the time lag between reducing obesity levels and the impact on extending life spans, most healthcare providers would argue that near-term savings are a more important factor than any concerns over the future costs associated with age extension. Similar to other prevention topics, evidence would suggest that programmes targeting entire populations are less cost effective than more innovative initiatives based upon understanding the motivation and behaviour of different consumer groups (Murrow and Welch, 1997).

References

Bauer, C., Herz, B. and Karb, V. (2003), The other twins: currency and debt crises, *Jahrbuch für Wirtschaftswissenschaften*, Vol. 54, No. 3, pp. 248–268.

Borins, S. (2002), Leadership and innovation in the public sector, *Leadership & Organization Development Journal*, Vol. 23, No. 8, pp. 467–476.

Brooker, P. (2005), *Leadership in Democracy: From Adaptive Response to Entrepreneurial Initiative*, Palgrave Macmillan, Basingstoke.

Chaston, I. (2011), *Public Sector Management: Mission Impossible?* Palgrave Macmillan, Basingstoke.

Currie, G., Humphreys, M., Ucbasaran, D. and McManus, M. (2008), Entrepreneurial leadership in the English public sector: paradox or possibility?, *Public Administration*, Vol. 86, No. 4, pp. 987–1008.
DeLeon, L. and Denhardt, R.B. (2000), The political theory of reinvention, *Public Administration Review*, Vol. 60, No. 2, pp. 89–97.
Dunleavy, P., Margetts, H., Bastow, S. and Tinkler, J. (2005), New Public Management is dead—long live Digital-Era Governance, *Journal of Public Administration Research and Theory*, Vol. 16, pp. 467–494.
Goldfinch, S. (2006), Pessimism, computer failure, and information systems development in the public sector, *Public Administration Review*, Vol. 67, No. 5, pp. 917–992.
Herps, J.M., Van Johannes, H.H., Halman, I.M., Martens, H.M. and Borsboom, R.H. (2003), The process of selecting technology development projects: a practical framework, *Management Research News*, Vol. 26, No. 8, pp. 1–16.
Hoenig, T.M. (2008), Maintaining stability in a changing financial system: some lessons relearned again?, *Economic Review—Federal Reserve Bank of Kansas City*, Vol. 93, No. 1, pp. 5–18.
Husig, S. and Mann, H. (2010), The role of promoters in effecting innovation in higher education institutions, *Innovation: Management, Policy & Practice*, Vol. 12, pp. 180–191.
Kaufman, G.G. and Wallinson, P.J. (2001), The new safety net, *Regulation*, Vol. 24, No. 2, pp. 28–36.
Leutwyler, K. (1995), The price of prevention, *Scientific American*, Vol. 111, No. 4, pp. 124–129.
Lindbeck, A. (1995), Hazardous welfare-state dynamics, *The American Economic Review*, Vol. 85, No. 2, pp. 90–102.
Lines, R. (2007), Using power to install strategy: the relationships between expert power, position power, influence tactics and implementation success, *Journal of Change Management*, Vol. 7, No. 2, pp. 143–170.
Moore, M.H. (1995), *Creating Public Value: Strategic Management in Government*, Harvard University Press, Boston, MA.
Morris, M.H. and Jones, F.F. (1999), Entrepreneurship in established organisations: the case of the public sector, *Entrepreneurship Theory & Practice*, Vol. 1, pp. 71–91.
Murrow, J.L. and Welch, J. (1997), Improving marketing strategies for wellness, *Marketing Health Services*, Summer, pp. 30–39.
Nesvetailova, A. (2005), United in debt: towards a global crisis of debt-driven finance, *Science & Society*, Vol. 69, No. 3, pp. 396–419.
Pandey, S.K. (2010), Cutback management and the paradox of publicness, *Public Administration Review*, July/August, pp. 542–556.
Parnaa, O. and von Tunzelmann, N. (2007), Innovation in the public sector: key features influencing the development and implementation of technologically innovative public sector services in the UK, Denmark, Finland and Estonia, *Information Policy*, Vol. 12, pp. 109–125.
Rappange, D.R., Brouwer, W.B., Hoogenveen, R. and Van Baal, P. (2009), Healthcare costs and obesity prevention: drug costs and other sector-specific consequences, *Pharmacoeconomics*, Vol. 27, No. 12, pp. 1031–1044.
Reger, R., Mullane, J., Gustafson, L. and DeMarie, S. (1994), Creating earthquakes to change organizational mindsets, *The Academic of Management Executive*, Vol. 8, No. 4, pp. 31–43.
Stanaland, A.J.S. and Gelb, B.D. (1995), Can prevention be marketed profitably, *Journal of Health Care Marketing*, Vol. 15, No. 4, pp. 59–68.

Schettkat, R. (2001), How bad are welfare-state institutions for economic development?, *Challenge*, Vol. 44, No. 1, pp. 34–55.

Teutsch, S.M. (1992), A framework for assessing the effectiveness of disease and injury prevention, *Morbidity and Mortality World*, Vol. 41, No. 3, pp. 1–12.

Weiner, S.E. (1995), Budget deficits and debt: a summary of the bank's 1995 symposium, *Economic Review—Federal Reserve Bank of Kansas City*, Kansas City, Vol. 80, No. 4, pp. 5–18.

Wolf, M. (2011), We should declare an end to our disastrous war on drugs, *Financial Times*, London, June 4th, p. 11.

Wright, B.E. (2007), Public service and motivation: Does mission matter?, *Public Administration Review*, Vol. 67, No. 1, pp. 54–64.

15
Leaner Futures

Prosperity trends

Over the last two decades the poor economic performance of many Western nations has already resulted in people facing a decline in average disposable income and a consequent reduction in their standard of living. In the UK and the USA the global financial crisis impacted house prices leading to a further reduction in personal wealth by over $700 billion and $9 trillion respectively (Anon, 2012a). Even more pain can be expected in the future as governments seek ways of reducing excessive public sector deficits. Activities associated with rebuilding national balance sheets will result in many people having to pay more taxes and concurrently self fund services which in the past were available through a now rapidly contracting welfare state. Ultimately a nation's economic revival is determined by the export performance, which in turn increases the spending power of the populations. Unfortunately few Western nations can expect any near-term significant growth in total exports. This is because any gains in sales to the emerging nations will be offset by poorer demand in domestic markets and overseas markets in Europe and North America.

With limited economic growth through exports and declining domestic consumer disposable incomes, the outlook for most Western world companies is that survival over the next few years will remain extremely difficult. Eventually some economies can be expected to start showing signs of recovery. However the duration of time which will pass before this occurs is difficult to predict and will vary dramatically between nations. For example, the US economy is beginning to show some 'green shoots' of recovery in 2012. At the other extreme countries such as Greece are expected to continue to face economic austerity until well into the next decade.

Organisations seeking to identify strategies to sustain performance should recognise two key factors that can influence future outcomes. Firstly as identified by Schumpeter (1934) during the Great Depression, the businesses which emerge stronger and more successful at the end of a prolonged economic downturn are those that focus upon adopting an entrepreneurial orientation and engage in the development of innovative new products, services or operational process. Secondly there are some segments of markets and certain industries which can expect to be less adversely impacted by a lengthy period of austerity. These include sectors supplying necessities such as basic foodstuffs, energy and healthcare services.

Due to the nature of their internal competences, some organisations will be unable to move away from their existing core markets where overall growth prospects will remain very limited. The strategic focus in these cases must be about gaining a detailed understanding of changing customer attitudes and behaviours and then utilising this knowledge to exploit innovation as the basis for repositioning the organisation offering superior value versus competition (Raggio and Leone, 2009). An example of this approach is provided by Land Rover's marketing strategy in the USA. The brand has been positioned as offering an 'intelligent choice', based on the fact that, although more expensive than other vehicles, the product lasts longer, requires less maintenance and has a higher resale value than other 4x4 vehicles. Thus the benefit offered to consumers facing spending constraints is that purchasing a Land Rover will lengthen the time before purchase of the next replacement vehicle is necessary; thereby permitting amortisation of the Land Rover purchase over a longer time period.

Non-sustainable sustainable energy

Case Aims: To illustrate some of the problems associated with attempting to make greater use of sustainable energy.

Recent discoveries of new offshore oil and gas deposits around the world, accompanied by growing recognition of the new opportunities offered by exploiting shale gas resources, does mean the world is unlikely to run out of hydrocarbon fuel any time soon. Unfortunately the human race's apparently insatiable desire to own cars and electrical appliances means that demand will continue to rise. Consequently although new hydrocarbon resources will be coming on-stream, consumers can expect to face rising energy costs.

The desire to reduce greenhouse gas emissions has caused Western governments to stimulate a switch to renewable energy sources with emphasis on solar and wind technology. As governments in Europe have sought to control public sector spending, many have come to recognise that the subsidies to consumers or to solar panel installers have been excessively generous. This has resulted in governments in countries such as Germany, Spain and the UK subsequently reducing the level of subsidies. Governments had hoped these subsidies would accelerate an expansion in the size of their domestic renewable energy manufacturing sector. Unfortunately because much of the solar technology had already been transferred to China through joint ventures and alliances, plus the Chinese government's ongoing financial support for the industry, Western solar manufacturers have found increasing difficulties in sustaining operations. Many have cut back their expansion plans and in some cases have been forced into bankruptcy. The largest bankruptcy has been the company Solyndra, a solar panel manufacturer in California which in 2009 had secured $535 million in loan guarantees from the US government. This company's failure along with two of America's other largest industry members has left China as the world's dominant solar industry manufacturer, owning almost 60 percent of the global production capacity. Industry analysts attribute this situation to the fact that Chinese solar producers have been particularly adept at using low-cost loans from state-owned banks and other subsidies to undercut prices charged by companies in other countries.

The other renewable sector facing problems is wind energy. The technological leadership enjoyed by Western manufacturers has been eroded by growing capability in China, which has led to the closure or downsizing of Western turbine manufacturers. However concerns over the noise and visual impact of land-based wind turbines have caused some Western governments to switch their support to off-shore, large scale wind farms. The technological problems associated with operating off-shore have meant that to date certain Western manufacturers have retained their leadership position relative to emerging nation competitors. The remaining, and probably larger, problem facing the industry is that the utility companies are reliant upon feed-in-tariffs to cover the higher costs of operating wind turbines. Consequently consumers are facing higher energy bills and politicians are under increasing pressure to reduce these feed-in tariffs. Should this occur then the result will be fewer new wind farms being brought on-stream in the future.

Consumer markets

Over the past 60 years, the traditional target for Western fmcgs and national retail chains has been families in the 18–45 year age group with children. This group is facing the sharpest fall in disposable income during the current downturn and as these households strive to survive, branded products will be adversely impacted as consumers switch to private label products. Dibadj et al (2010) examined purchase patterns across a range of grocery products in the USA following the onset of the current downturn.

A preference for lower prices was evident for certain value tier brands over the premium price alternative (e.g. Arm & Hammer laundry detergents versus Proctor and Gamble's Tide; Luvs disposable diapers versus Pampers or Huggies). Two thirds of consumers have already traded down. The categories most exposed to down trading are commoditised products such as bleach, plastic bags, bottled water and household cleaners. An additional concern for major brands in sectors such as diapers, trash bags, consumer tissue products, batteries, cat litter and beauty products is that American consumers apparently have no intention of trading back up when their personal financial circumstances improve. Furthermore 70 percent of respondents who have yet to down trade stated this would occur immediately should they become unemployed, experience a salary reduction or the overall economic situation continues to deteriorate.

Lamey et al's (2007) study of private label trends vitiated conventional wisdom that a country's own label share increases when the economy downturns and then decreases during an economic upturn. The researchers concluded that share gains enjoyed by private label products during today's downturns are now permanent because consumers are being persuaded that private label goods are a more sensible option no matter their personal financial circumstances. Given the current downturn is expected to be the longest since World War II, then many branded goods manufacturers can expect not just to suffer a major sales decline during the period of austerity, but even once an economic upturn begins a certain proportion of loyal customers will have been lost forever.

Two years into the current downturn the market research company Nielsen estimated in the UK that by the end of 2012 branded goods' share of grocery purchases will have declined from 70 to 57 percent (Harper, 2010). The company noted however that trends for brands will vary by category. Tobacco and beer brands appear to be able to sustain customer loyalty. In the soft drinks category there has been a slight improvement in loyalty. This is not the case for products such as laundry detergents and household cleaning products where consumers appear very willing to switch to either cheaper brands or a store's private label offering.

Hillier and Baxter (2001) concluded that the probability that major brands are more likely to survive an economic downturn is increased when they engage in higher expenditure on new product development. These researchers determined that decreased emphasis on developing new or improved products will erode a brand's ability to combat a permanent loss of market share to private label alternatives. One company in the UK which has demonstrated the benefits of proactive marketing involving sustaining the development and launch of new products during previous recessions is Reckitt Benckiser (Anon, 2005). Even when overall economic conditions have adversely impacted total consumer spending, the company has achieved profit growth in large part by relying upon new products which, when launched, are supported by higher advertising spend than the competition.

Thrift and opportunity

Case Aims: To illustrate that even in a prolonged economic downturn opportunities do exist to generate higher revenue.

Even in an economic downturn consumers will still purchase products and services although in many cases the selected items may be very different from goods purchased during better economic times. An example of this scenario in the UK is provided by the nation's charity shops. With UK consumers finding it increasingly difficult to 'make ends meet' in the current period of prolonged austerity, a major beneficiary of changing buyer behaviour has been the country's charity shops. According to the Association of Charity Shops, turnover grew by 2.4 percent between April and June 2010 and profits rose by 9 percent (Barrett, 2010). The increasing numbers of retailers being forced into bankruptcy or needing to close some of their outlets has led to the number of vacant retail sites available to charity shops rising by 30 percent over the last three years.

An example of success is provided by Banardo's, a charity which focuses upon helping children through assisting adoptions and operating orphanages. The Banardo's shops achieved a profit of £5.7 million in 2009/10 and expected to achieve £7 million in 2010/11. This compares with a profit of only £1.7 million in 2005/06. The charity has also begun to invest in improving their retail marketing activities and already has opened 17 new stores which will specialise in the supply of children's clothes. The organisation has also launched a second brand, Brand 2, which is the name the charity has adopted for merchandising higher quality, designer goods that can be sold at a premium price.

Another sector enjoying revenue growth are pawnbrokers (Wallace, 2009). H&T is the UK's largest pawnbroker and in 2009 their annual profits increased by 52 percent. There has not been a huge upsurge in new customers but existing customers are using H&T facilities more frequently both to raise money and to borrow funds for short periods of time. The chain deals primarily in jewellery. Most customers just want a quick cash injection to cover immediate bills and the average loan is in the region of £130. Similar growth has been enjoyed by pawnbrokers in the USA (Rivlin, 2011). There the vast majority of customers are seeking a short-term loan and pledge an item such as a watch or a wedding ring as collateral. The amount borrowed is typically small, in the region of $100. Four out of five customers successfully pay off their loan and retrieve the item which has been placed with the pawnbroker. Nevertheless this approach to borrowing money is expensive. Fees charged equate to an annual interest rate of between 50 and 250 percent.

Older consumers

Older middle-class consumers in Western nations often referred to as 'baby boomers' tend to be less impacted by austerity. This is because many have accumulated assets and pensions as a result of being in work for many years prior to the onset of the 2009 global recession. Further insights on the opportunities in the baby boomer market can be gained from the forecast in Table 15.1 of the spending patterns of older people in the US market in the year 2015 (Court et al, 2007).

Companies deciding to focus more efforts on building market share among older members of a population do, however need to avoid perceiving this as a homogeneous market containing individuals with similar needs and requirements. The nature of the differences can be illustrated by market segmentation studies concerning older people. Moschis (1992, 1996) utilised gerontographics, which involved a large-scale survey of retirees in the USA covering 136 measures of exhibited characteristics such as biophysical, psychological and ageing. From these data Moschis used statistical analysis to classify respondents into four basic types. These are *healthy indulgers* (focus is on enjoying life), *healthy hermits* (healthy but socially withdrawn), *ailing outgoers* (still active despite declining health) and *frail recluses* (chronically ill, often living in isolation).

Table 15.1 Spending patterns of older people in the US market

Total Market Share of US Markets By 50+ Households In 2015	
Sector	% Market Share
Food at home	52%
Food away from home	50%
Health and beauty aids	60%
Housewares	58%
Furniture and appliances	52%
Clothing	48%
Footwear, other apparel items	44%
Consumer electronics	50%
Medical services	74%
Prescription drugs	83%

US Net Worth and Expenditure By Age Group 2015			
	Net Worth	Consumption	Income
Born before 1945	22%	11%	11%
Boomers	57%	40%	41%
Gen X (Born 65–81)	20%	38%	38%
Born After 1982	1%	11%	11%

Dr Ken Dychtwald (2005) has also undertaken research that lead to classifying retirees into distinct customer segments. From his studies he concluded that retirees in America could be divided into four segments. The *Ageless Explorers*, who make up 27 percent of retirees and have carefully planned their retirement to permit them to remain active, productive and independent in retirement. The *Comfortably Contents*, who constitute 19 percent of retirees and have also planned for retirement but lead a more relaxed lifestyle with emphasis on travel and recreation. The *Live For Todays*, 22 percent of retirees who are financially unprepared and, although aspire to be Ageless Explorers, lack the financial resources required to support such a lifestyle. Finally the *Sick & Tireds*, who constitute 32 percent of retirees and are often widowed or in poor health; thereby causing them to be resigned to a somewhat unfulfilled and less than satisfying lifestyle.

Another issue is the role which older people are currently fulfilling. Schewe and Balzs (1992), for example, proposed that the following will have significant influence over an individual's social behaviour and consumption patterns:

(1) *The Empty Nester Role*, which within the family life-cycle occurs when the children depart the home leaving the parent(s) freer to fulfil their own personal needs and desires.
(2) *The Care Giver Role*, which is becoming increasingly common as people's parents are living longer and, because of infirmity, the children need to become the parent. With young families facing the increasing financial stress of attempting to earn sufficient money to live, older people are taking on the care-giver role of providing economic support to their children.
(3) *The Retirement Role*, where either one or both individuals retire, leaving them free to participate in activities totally of their own choice such as a hobby, returning to education or spending more time playing sport.
(4) *The Grandparent Role*, which, because older people are living longer and remaining healthier, means people are becoming much more actively involved in entertaining their grandchildren and fulfilling the role of being a alternative source of guidance and support.
(5) *The Widowhood Role*, which is caused by women having a longer life expectancy than men with widowers in Western nations outnumbering men by approximately five to one—and past the age of 65, approximately 50 percent of pensioners are widows. Widows 'feeling younger' until later in life means there is a trend towards these ladies being much more socially active and engaging in the consumption of luxury goods.

Sustaining success

The Great Recession, subsequent problems associated with reducing public sector deficits and the threat associated of America loosing economist leadership to China have been described by Rachman (2010) as the causes for the 'The Age of Anxiety'. Influencing this anxiety is the fact that repairing the economic damage done by past excessive consumer and government borrowing in some Western nations will result in a prolonged period of austerity: in some such as Greece and Spain this is expected to last until well into the next decade.

This will (i) dampen demand for the output from domestic industry, (ii) increase level of low-cost imports from the emerging nations, thereby further reducing sales by domestic firms, and (iii) further increase unemployment leading to an ongoing further decline in the average standard of living.

Sanders (2007) concluded the only really certain effective survival strategy for organisations based in developed economies is to focus on the exploitation of new technologies as the basis for supporting 'radical innovation' leading to the creation of new products or services or, in some cases, the establishment of entirely new sectors of industry. A similar view concerning radical innovation has been expressed by Tellis et al (2009, p. 3) who posited that:

> Radical innovation is crucial to the growth of firms and economies. It merges some markets, creates new ones, and destroys old ones. It can propel small outsiders into a position of industry leadership and can bring down large incumbents that fail to innovate. Firms at the leading edge of radical innovation tend to dominate world markets and to promote the international competitiveness of their home economies.

In the second half of the 20th century, global economic growth and rising standards of living permitted firms to exploit innovation as the basis for enhancing consumer lifestyles. A combination of factors including ongoing increases in energy costs, population ageing and the non-sustainability of welfare states mean that companies attempting to build market share by promising consumers further improvements in lifestyle is no longer a viable strategy. Instead the focus on successful innovation for the foreseeable future will be that of developing new propositions that enhance affordability. Organisations have the options of delivering the improved affordability through price or superior performance. In most cases the latter option is more likely to protect a Western firm from competition, especially in relation to competitors based in the emerging nations who are intent on building global market share by offering lower priced goods.

Business model innovation

Coulston-Thomas (2005) concluded that business losers tend to be indecisive and oblivious to the need for change. Their response is reactive, reflecting an inability to respond to market signals that environments are likely to be very different in the future. Losers are usually led by indifferent managers who confuse operational and strategic issues and fail to provide their organisations with a compelling vision or clear direction. Imitation and following others is preferred over independent thought. There is also a tendency to be short-term thinkers and resistant to new ideas from external sources.

In contrast winners accept the need to regularly re-invent their operations and are committed to exploiting innovation to achieve long-term goals. Their managers are able to motivate staff through inspiration and by providing clear direction over organisational priorities. Senior management are open to new ideas. Calculated risk taking is encouraged and great importance is attached to stimulating learning across the entire work force. Proactive response is always preferred over reacting to events, accompanied by a willingness to push back the boundaries of what is considered feasible or achievable.

In an increasingly competitive world where the speed of change has been accelerated by the advent of new technologies, some organisations are achieving success through engaging in business model innovation (BMI). Three strategic opportunities offered by BMI (Shelton, 2009) are:

(1) *Target Customer Revision*, which involves focusing upon a completely different target customer group(s).
(2) *Value Proposition Revision*, which involves changing the nature of the value proposition of the product or service.
(3) *Value Chain Revision*, which involves making changes to how a product or service is created to enhance the value chain both within and outside of the organisation.

In terms of implementing BMI, Koen et al (2011) suggested the two key dimensions are technology and market system participation. In the case of technology, the options available to the innovator are incremental, architectural or radical technologies. Incremental projects can be managed using existing capabilities and knowledge. In contrast architectural and radical technology projects often require very different knowledge sets. Architectural innovation involves creating new ways to integrate components into a system based on current or incremental changes to existing technology. Radical innovation involves exploitation of an entirely new core technology.

The market system dimension encompasses the organisation's interaction with other participants within a supply chain. Revising relationships within the system is often the basis for creating a new value network. Within BMI the value network dimension can be divided into various innovation opportunity options. These include innovations within the company's existing network, innovations requiring networks with elements that are new to the company, networks directed towards existing customers, networks directed to non-users in existing markets and networks targeting non-customers in totally new markets. Implementation of BMI may also involve disruptive innovation.

Christensen et al (2004) suggested that disruptive innovation can be divided into two types; namely 'low end' and 'new market' disruption. Low end disruption generates market share through offering a lower price proposition to existing customers. By enhancing affordability, current customer loyalty can be retained and non-users in a current market may be converted into becoming users. New market disruption involves the organisation focusing efforts on developing a completely new customer group.

Koen et al (2010) identified the following potential obstacles to successful BMI:

(1) *Paradoxical Leadership*, reflecting a tendency of leaders to do the wrong things for the right reasons, such as applying existing knowledge to solve totally new problems.
(2) *Organisational Complexity*, in terms of selecting an inappropriate organisational structure for implementing BMI.
(3) *Innovation Management*, in terms of selecting the wrong organisational structure.
(4) *Financial Uncertainty*, created because existing forecasting models may be inapplicable to the revenue and cost structures associated with a proposed BMI.
(5) *Prior Knowledge Barriers*, caused by implementers using existing knowledge when what is required is the acquisition of totally new knowledge, often completely outside of the organisation's areas of competence.

An evolving business model

Case Aims: To illustrate that in high-technology sectors business models may need to be revised after only a few years in use.

The capability to map the human genome has provided scientists and clinicians with new, more powerful tools to study the role that genetic factors play in much more complex diseases, such as cancer, diabetes and cardiovascular disease, which constitute the majority of health problems in the developed nations. Although the genomics sector has existed for approximately 20 years, the industry has moved through a number of business model innovations in a search for sustainable commercial viability (Rothman and Kraft, 2006). The aim of most genomics start-ups centred upon the development of proprietary technology or knowledge protected by patents. The high cost of R&D accompanied by strict government regulations over the approval of new healthcare treatments subsequently forced most of these firms to move to a business model based upon forming alliances and joint ventures with large pharmaceutical companies. In the 1990s the large drugs firms sought to use expertise within genomics companies to act as a source of drug innovations through the development of genomic-based drugs. This drug discovery business model proved to be extremely short lived. Firstly rapid development of lower cost DNA sequencing technologies reduced this area of knowledge generation to that of a being a virtual commodity good. Secondly the development of automated generation and screening systems which accelerated the speed with which the new genomes could be researched were adopted by the big pharmaceutical firms and rendered the smaller firms' upstream analysis role redundant.

In response to these industry changes those genomics companies that have survived have been forced to again re-define their business model. In most cases this has involved a move downstream in the healthcare supply chain by becoming engaged in the actual development of new drugs. This new business model requires much larger expenditure on R&D before a saleable product is created, which in turn increases the level of business risk. The advantage however is that once a product has successfully passed through pre-clinical trials that product has much higher commercial value than a new compound supplied as an ingredient to the major drugs manufacturers. This latest business model revision has demanded the acquisition of new knowledge competences. Companies have achieved this through a mix of mergers, acquisitions, joint ventures and alliances with downstream pharmaceutical firms.

Smart upwave

Currently the most likely source of success through innovations is in computing, electronic communications and exploitation of the Internet (Wright and Dawood, 2009). Tyson (1998) posited that the scale of benefits of computers and related technology would rise more rapidly once certain technological advances permitted the world to enter the

'Intelligence Age'. His predictions over these required technological advances included:

(1) The convergence of electronic devices such as computers, telephones and televisions to permit all of these to be linked via the Internet.
(2) Faster download speeds and the ability to link data acquisition systems such as remote sensors sending data via the Internet without the need for hard-wired connections.
(3) Access to significantly larger lower cost data storage and analysis capability, which is now available following the arrival of cloud computing.
(4) More sophisticated data analysis through access to low-cost advanced statistical analysis software of the type used in automated customer relationship management (CRM) systems.
(5) The automation of decision making within organisations by exploiting advances in artificial intelligence (AI) systems.

All of these advances are now in place. This outcome is seen by many experts and futurists as the basis of the next global economic upwave, which has been being labelled the 'Smart Age' (Anon, 2010). The new era offers Western firms three very significant opportunities. Firstly the leading edge competences in computing and IT still reside in Western companies such as Apple, Google, IBM, Intel and Oracle. Secondly Western firms have superior core competence permitting exploitation of smart age technology to sustain existing competitive advantage. Thirdly smart technology permits the creation of new solutions to world problems such as the energy crisis, global warming, population ageing, stabilising healthcare costs and reducing the costs of delivering welfare state services.

Bughin et al (2010, p. 28), in their assessment of the implications of the Smart Age, concluded that exploitation of new opportunities provided a vast array of opportunities for entrepreneurs. In support of this assessment they noted that:

> The pace of technology and business change will only accelerate, and the impact of the trends below will broaden and deepen. For some organisations, they will unlock significant competitive advantages; for others, dealing with the disruption they bring will be a major challenge. Our broad message is that organisations should incorporate an understanding of the trends into their strategic thinking to help identify new market opportunities, invent new ways of doing business, and compete with an ever-growing number of innovative rivals.

As summarised in Figure 15.1, the arrival of the Smart Age means organisations need to determine whether they have, or need to develop, internal

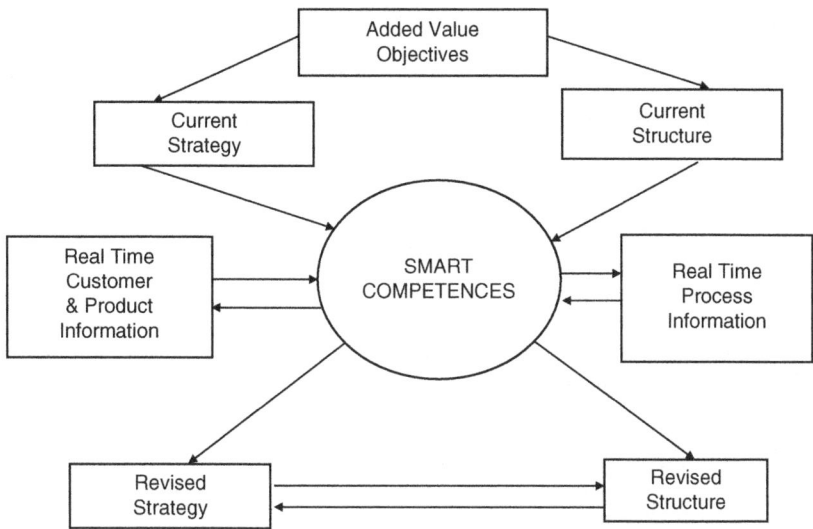

Figure 15.1 Identifying smart strategy and structure

competences capable of exploiting the new sources of opportunity to enhance product or service provision affordability. One source of opportunity is the acquisition of real time data about customers, which permits rapid identification about how new or improved products or services can be made available and also supports more accurate customer targeting. The second opportunity lies in those smart products which offer new services to customers while concurrently sustaining performance of existing products. The third opportunity is that organisations are able to access real time data about all aspects of internal organisational processes. This assists the rapid identification of how processes might be improved and provides early warning of developing performance problems before these have an adverse impact of the organisation.

The objective of smart technology is often to replace humans with some form of non-human system capable of acquiring, processing and analysing data. In relation to the exploitation of smart technology the outcome will be of the following types:

(1) *Tangible applications,* where the technology is used to enhance the customer's product or service experience through either improving performance or reducing costs (e.g. making cars safer by equipping vehicles with sensors that automatically initiate evasive action when an accident is about to happen).

(2) *Information applications*, where the technology is used to enhance information available to the user (e.g. available supplier options identified by visiting a price comparison website).
(3) *Layered applications*, where the technology is used to direct the user to another knowledge source or assists in supporting a superior purchase outcome (e.g. search engines such as Google and Yahoo).

Sustained smart wave innovation

Case Aims: To illustrate how the smart upwave provides new opportunities for business growth through innovation.

An excellent exemplar of how the Internet and associated convergent electronic technologies have provided entrepreneurs with a vast array of opportunities for sustaining innovation and developing new business models is provided by the activities of founder of Amazon.com, Jeff Bezos (Anon, 2012b). His initial entry into online marketing occurred at a time when the Internet was still at a very early stage of development and many of the components of the required technology had to be developed by the company. The required investment and the decision to engage in the retail mass marketing of books where unit profit margins are low meant it was several years before this first venture made a profit.

As the company acquired expertise in areas such as back office operations and the management of large data sets, Bezos recognised that this knowledge permitted Amazon to begin generating revenue by assisting others to move into online marketing. Amazon Web Services are now used by thousands of firms ranging from small start-ups through to major corporations such as the Swedish telecoms giant Ericsson. Bezos was also one of the pioneers in the world of cloud computing and, again, as knowledge was acquired this was exploited by the company—now a leader in the new world of supplying cloud-computing services to other organisations.

Bezos also recognised the opportunities associated with using tablet computers to create the firm's e-reading device, Kindle. He insisted that the product must be able to be used without the requirement to being plugged into a PC in order to download materials. He also wanted the product to work anywhere, not merely in WiFi hotspots. The solution was to devise a new business model whereby users could download e-books via their mobile phone without incurring network usage fees. The success of Kindle, originally launched in 2007, has now been followed by Kindle Fire which has all of the features of an advanced

> tablet computer; thereby permitting customers to use the device to access Amazon's apps, video streaming and music download services. The major advantage over most other suppliers of tablet computers is that Amazon can offer Kindle Fire at a highly competitive price by cross-subsidising the product through the sale of the company's huge range of digital content.

Healthcare affordability

One of the opportunities for enhancing healthcare is utilisation of smart technology's capability to manage and exploit the data which is generated during the diagnosis and treatment of patients. Known as 'm-health' this activity has been greatly assisted by the advent of mobile technology. It can be expected that effectiveness of m-health will continue to be enhanced as information and telecommunication infrastructures converge to create new mobile health systems. This is because mobile technology can offer benefits in the areas of availability, miniaturisation, speed and communication bandwidth (Simpson, 2003).

By utilising wireless-based computing, healthcare providers can access, receive, update and transmit critical patient and treatment information. Computer-based documentation of care will assist in eliminating the human error that can occur during the recording and storage of patient records. Additional benefits will accrue from the use of picture archiving and communication systems (PACS) which permit the computerisation of data such as radiologic film. Users can acquire, store, transmit and display images digitally, which permits linking intensive care units in hospitals with medical staff based in other locations.

Computerised provider order entry (CPOE) systems are capable of identifying, and thereby preventing, potential medical errors at the earliest possible point in the treatment process. The healthcare provider enters a proposed action into the computer and the automated system checks for possible causes of error such as incorrect dosages, wrong drug, drug-allergy interactions and drug-to-drug and drug-food allergies. This technology removes the guesswork from medication administration, reduces staff stress levels, improves clinical effectiveness and enhances productivity by allowing staff to focus on patient care rather than paperwork. Even greater affordability opportunities are being offered by new technologies that allow organs and bodily functions to be permanently monitored by doctors even when the patient has returned home after surgery. Bio-analytical micro-systems can also be used for activities such as determining blood sugar concentration and providing guidance to the patient over self-treating conditions such as diabetes.

Smart implants

Case Aims: To illustrate how healthcare treatments can be enhanced by exploiting advances in smart technology.

In the past healthcare treatments of conditions such as heart disease have relied upon early and often continual diagnosis to be effective when treating the patient. Recent developments in the miniaturisation of sensors are beginning to now render such approaches obsolete. Sensors have been developed which the patient can wear or have implanted into their body that monitor variables such as blood pressure or provide electrocardiograms (ECG). Once these tiny instruments register an irregularity, they automatically contact a diagnosis centre where doctors are on call on a 24/7 basis (Anon, 2008). The potential of these sensors has, in recent years, been greatly enhanced by mobile devices such as smartphones, which can provide healthcare staff with patient data located at home or at work. By linking these data streams to algorithms that assess risk or detect early warning signs, healthcare providers are in a much better position to implement earlier intervention and thereby halt further deterioration in a patient's medical condition (Murray, 2012). The advent of smartphones and sensors is also assisting people to engage in self tracking and self management of medical conditions (Anon, 2012c).

Intelligent implants are seen as a critical aspect of future medical treatments. For instance, the risk when fitting a new hip joint is that the joint could be rejected by the surrounding tissue. Intelligent implants measure how the patient's immune system is reacting and transmit data via radio signal to the doctor in order that the problem can be localised and immediately treated. Implanted sensors can also be used to monitor bone healing after prosthetic implant surgery. For example, after a knee or hip replacement the patient is physically restricted in order to provide a stable environment for their bones to grow around the implant. Strain gauges fitted to the bone can measure the elongation and send data to an external data storage system permitting the doctor to monitor the progress of the healing process and be provided with early warning of post-surgical problems as these arise (Connolly, 2009).

References

Anon (2005), Craziness makes merger work for Reckitt Benckiser, *Financial Times*, London, February 9th, p. 21.

Anon (2008), German medical implant technology makes patients more independent, *German Business Review*, January, p. 5.

Anon (2010), It's a smart world, *The Economist*, London, November 6th, pp. 3–24.
Anon (2012a), Lost economic time: the Proust index, *The Economist*, London, February 25th, pp. 73–74.
Anon (2012b), Taking the long view, *The Economist*, London March 3rd, pp. 21–22.
Anon (2012c), Counting every moment, *The Economist*, London March 3rd, pp. 19–20.
Barrett, H. (2010), Bucking the trend on the high street, *Third Sector*, London, November 9th, pp. 14–15.
Bughin, J., Chui, M. and Manyika, J. (2010), Clouds, big data, and smart assets: ten tech-enabled business trends to watch, *McKinsey Quarterly*, Issue 4, pp. 26–43.
Christensen, C.M., Anthony, S., and Roth, E. (2004), *Seeing What's Next*, Harvard Business School Press, Boston, MA.
Connolly, C. (2009), Sensor trends in processing and packaging of foods and pharmaceuticals, *Sensor Review*, Vol. 27, No. 2, pp. 103–111.
Coulston-Thomas, C. (2005), Corporate transformation, *Management Services*, Summer, pp. 14–16.
Court, D., Farrell, D. and Forsyth, J.E. (2007), Serving ageing baby boomers, *The McKinsey Quarterly*, Issue 4, pp. 102–112.
Dibadj, A., Powers, S. and Keswani, P. (2010), US household and personal products: the Great Recession and shifts in consumer behaviour, *Black Book—U.S. Household & Personal Products*, Washington DC, March, pp. 1–27.
Dychtwald, K. (2005), Ageless ageing: the next era of retirement, *The Futurist*, Vol. 39, No. 4, pp. 16–22.
Harper, T. (2010), Choosing the right tools: brands, recession recovery, *Checkout*, London, August, pp. 60–61.
Hillier, T. and Baxter, M. (2001), How to prevent a hangover: brave strategies for difficult times, *Market Leader*, New York, Autumn, pp. 1–5.
Koen, P.A., Bertels, H.E.J., Elsum, I.R., Orroth, M. and Tollett, B.(2010), Breakthrough innovation dilemmas, *Research Technology Management*, November/December, pp. 47–55.
Koen, P.A., Bertels, H.E.J. and Elsum, I.R. (2011), The three faces of business model innovation, challenges facing established firms, *Research Technology Management*, June/July, pp. 56–65.
Lamey, L., Deleersnyder, B., Dekimpe, M.G. and Steenkamp, J.E.M. (2007), How business cycles contribute to private-label success: evidence from the United States and Europe, *Journal of Marketing*, Vol. 71, pp. 1–15.
Moschis, G.P. (1992), Geronographics: a scientific approach to analysing and targeting the mature market, *The Journal of Services Marketing*, Vol. 6, No. 3, pp. 17–27.
Moschis, G.P. (1996), Life stages in the mature market, *American Demographics*, Vol. 8, No. 9, pp. 44–49.
Murray, S. (2012), Remote care vital to reduce costs, *Financial Times*, London, February 27th, p. 3.
Rachman, G. (2010), *Zero-Sum World: Politics, Power and Prosperity after the Crash*, Atlantic Books, London.
Raggio, R.D. and Leone, R.P. (2009), Preserving and growing brand value in a downturn, *Journal of Brand Management*, Vol. 17, pp. 84–89.
Rivlin, G. (2011), It's a hot time to be a pawn star, *Newsweek*, New York, June 27th, pp. 20–21.
Rothman, H. and Kraft, A. (2006), Downstream and into deep biology: evolving business models in 'top tier' genomics companies, *Journal of Commercial Biotechnology*, Vol. 12, No. 2, pp. 86–98.

Sanders, M. (2007), Scientific paradigms, entrepreneurial opportunities and cycles in economic growth, *Small Business Economics*, Vol. 28, No. 4, pp. 339–355.

Schewe, C.D. and Balazs, A.L. (1992), Role transitions in older adults: a marketing opportunity, *Psychology & Marketing*, Vol. 9, No. 2, pp. 85–96.

Schumpeter, J. (1934), *The Theory of Economic Development*, Harvard University Press, Boston, MA.

Shelton, R. (2009), Integrating product and service, *Research Technology Management*, May/June, pp. 36–45.

Simpson, R.L. (2003), Today's challenges shape tomorrow's technology, part 2, *Nursing Management*, Vol. 34, No. 12, pp. 40–44.

Tellis, G.J., Prabhu, J.C. and Chandy, R.K. (2009), Radical innovation across Nations: the pre-eminence of corporate culture, *Journal of Marketing*, Vol. 73, pp. 3–23.

Tyson, K.M. (1998), Perpetual strategy: A 21st century essential, *Strategy & Leadership*, Vol. 26, No. 1, pp. 14–18.

Wallace, D. (2009), The pawnbroker's chain, *Management Today*, London, June, pp. 82–83.

Wright, C.S. and Dawood, I. (2009), Information technology: market success to succession, *The Review of Business Information Systems*, Vol. 13, No. 4, pp. 7–20.

Index

A
ABN Amro 41
absorptive capacity 154, 168, 172
accountancy firms 205
affluent consumers 28
affordability 250
affordable food 55
Airbus 109
Alcan 130–131
alliances 170
Allstate Insurance 169
Amazon 101–102, 184
ambidextrous 98
American consumers 70
Amtico 201
analysers 119
Android operating system 88
Apathetic Materialists 20
Apple 16, 65, 88, 142, 187
Applegate, A. 40
apps 178
Arla Foods 164
ARM Holdings 65
assessing futures 112–113
assurance 180
audacious goals 107
austerity marketing 28–39, 183
austerity 12, 29, 50, 71–72, 110, 115
austerity-proofing 204
Austrian School of Economics 45

B
B2B markets 69, 198
B2B organisations 192
B2B services 205
B2B value innovation 132–133
Baby Boomers 25
BAE Systems 62
Bandit Industries 146
bank lending 212
Bank of England 8
bank wars 38–42
banks 182
base technologies 147

BCG 143
Beechcraft 207
benefit innovation 118
benefit proposition 126
Bezos, J. 254
Billerud 164
bio-fuels 6
biosciences 203
biotech firms 151
Blue ocean strategy 121–122
Boeing 97, 108–110
Boliden 164
boundary spanning activity 139
brand consciousness 74
brand wars 37
branded goods 244
Braun Medical 193
breakthrough innovations 145
BRICs 13
budget purchasing 114
Build-A-Bear Workshop 122
Burger King 38
business model innovation 186, 249
business model upgrades 186

C
Canon 135
capitalism 20
car industry 94
Casio 158
Caterpillar Inc. 192
CDOs 8
Central Banks 3, 7, 225
charity shops 245
Charles Schwab 40
chasm theory 158
China 7, 13, 55, 61, 81, 195
Chinese government 15
Churchill, W. 104
classifying innovation 125
clock building 107
cloud computing 254
Coca Cola 37
collaboration 169–170

259

co-location 196
colonisation 20
comfortably well-off 75
comparative advantage 60–61
competence leadership 95
competence trap 94
competence 96–97
competences 87–89
competitive advantage 92, 94, 115, 118, 204
competitive dynamics 95
competitor knowledge 92
complementary knowledge 171
complex alliances 151
concept identification 149
conceptualising 99
Concorde 97
conflict 163
Confucianism 21
connective capacity 168
conservatism 23
conspicuous consumption 27, 74
consumer income 70
consumer markets 243–244
consumer values 28
consumptionism 26–27
contending forces model 87
conventions 163
core competence 88–89
Corning Glass 111–112, 193
coupon redemption 77
coupons 77–78
creative destruction 45
crime prevention 235
cross-licensing agreements 154
cultural capital 22
customer centric 38
customer efficiency 188
customer experience 188
customer interaction 177
customer knowledge 92
customer loyalty 28, 178
customer motivation 29
customer orientation 130
customer segments 247
customer values 19–20
Customised Affordability 117
Customised Performance 116
cutback response 226–227
CVS 179
cyber warfare 63

D

DEC 135
decentralisation 166
decoupling 195
de-engineering 47
defence industry 62–63
defenders 119
delegated authority 98
deliberate strategy 120
de-linking 101
Dell 133, 171
demand typologies 198
Depression Generation 25
derived demand 69
desorptive capacity 154, 168
development innovation 117
differential pricing 178
differentiation strategy 116
diffusion of innovation 158
digital media 77
discovery-driven approach 84
discretionary spending 71
disposable income 12–13, 57, 241
disruptive innovation 126, 133
domain-specific innovation 214
domestic market demand 61
dominant technology 95
Domino's Pizza 185
Dow Chemical 154
downstream aggressors 81
downstream complementors 153
Draghi, M. 9
drones 63
Drucker, P. 16
drug companies 58
DuPont 197
dynamic capabilities 94, 97
dynamic competence 94

E

early adoption phase 157
ECB 9–11
eco-efficiency 202
economic cycle 92
economic development 22
economic leadership 60
economic meta-event 31
economic recovery 196
economics 45
ecosystems 153
effective vision 110–111

effectiveness innovation 117
efficiency innovation 117
egalitarian commitment 23
electronic calculators 158
emergent strategy 120
emerging economies 14, 62, 191
emerging technology 148
empathy 180
employee values 162
Encyclopedia Britannica 112
energy consumption 33
energy costs 55–56
enhanced affordability 134
enhanced economy 134
enhanced simplicity 132
entrepreneurial behaviour 97
entrepreneurial intuition 214
entrepreneurial leverage 99–100
entrepreneurial orientation 46, 113, 125
entrepreneurial pathway 100
entrepreneurial strategy 115
entrepreneurial style 126
entrepreneurs 126
entrepreneurship 45–46
entrepreneurial marketing 31
environmental complexity 120
environmental orientation 87
environmentally friendly 28
equity funding 216
ethical consumerism 28
EU Commissioners 9
Euro zone 9
Euro 9
European Central Bank (ECB) 226
European Union (EU) 211
experiential shopper 32
experimentation 99–100
exploiting innovation 142
extended innovation 169
external environment 118

F
Facebook 77
factor conditions 61
Family Expenditure Survey 70
FAO 55
Festo AG 147
fibre optics 111
first movers 62
flexibility 118–119
float glass 95

fmcgs 37, 57
focal firm 153
food industry 155, 198
food supplies 6
Ford 72
founder funding 216
fracking 5, 56
freemium model 48
Friedman, M. 3
Fukushima 7
functional quality 180
futuring 83–84

G
gastric pacemaker 85
GE Corporation 14,193
GE 193
General Motors 72
Generation X 25
Generation Y 26
Generation Z 26
generational values 24–25
generic drugs 58
Germany 9, 194
Gerstner, L. 44
Glass-Steagall Act 225
global economic growth 248
global economy 69
global population 55
global warming 6–7
globalisation 60
going green 201–202
Goodwin, F., Sir 41–42
Google 16, 88, 100, 142
Great Depression 1, 16, 32, 59, 225
Great Recession 27, 32, 248
Greece 9–11, 33, 226
green innovation 202
green products 48, 202
greenhouse gases 6–7, 242
Groupon 77–78, 180
growth strategies 61

H
Hayek, von, F. 3–4, 8
HBOS 41
healthcare affordability 255
healthcare budgets 58
healthcare costs 57
healthcare prevention 235
healthcare strategies 236

healthcare 4–5, 57–58
heterogeneity 180
Hewlett Packard 158
high affordability strategy 115
high volume goods 201
higher level capabilities 89
Hofstede, G. 21
Home Depot 169
home markets 192
horizontal networks 215
Hornby, A. 41
human genome 250–251
hybrid marketing 207
hybrid services 207
Hyundai 71

I
IBM 21, 31, 134, 196
IKEA 184
image 180
IMF 9–11, 226
immigration 20
immunisation 236
impulse purchasing 71, 114
India 61
individualism 22
industrial efficiency logic 131
Industrial Revolution 1, 16
inflation 2–3
informational promotion 28
innovation competence 90
innovation co-operation 170
innovation diagnostics 128
innovation explorers 138
innovation implementation 163
innovation marketers 138
innovation missionaries 138
innovation obstacles 127
innovation opportunities 94
innovation partnerships 50–51
innovation propositions 127
innovation strategies 91, 161
innovation 32, 44, 97, 194
inseparability 180
intangibility 180
intellectual autonomy 23
intellectual capital 91
Intellectual Property (IP) 90, 191
intellectual property vendors 66
Intelligence Age 252

inter-functional coordination 168
internal environment 98
Internet 188
inventive capacity 168
involuntary penny-pinchers 30
iPhone 143
IPM 42
iPod 195
Ireland 226
Italy 9

J
job creation 63
Jobs, S. 44–45

K
Keeler 204
key technologies 147
Keynes, J. M. 2
Keynesian economics 1–4
Keynesian policies 59, 63
Kindle 79, 195
kinship 19
Kirzner, I. 45–46
knowledge acquisition 94
knowledge complexity 167
knowledge economies 62
knowledge intensive logic 131
knowledge retention 168
knowledge 91
knowledge-based services 65–66
Kodak 90
Korea 22
KPIs 227
Kyoto Protocol 6

L
labour costs 61, 200
Land Rover 242
large firm competitors 218
lateral collaboration 170
leadership team 118
leadership 128
lean manufacturing 193
Learjet 207
legitimisation 163
Lehman Brothers 8
leveraging capabilities 100
Levitt, T. 78
lifestyle shifts 69

lifetime value (LTV) 42
live-for-today consumers 75
living standards 33
Lloyds-TSB 41
long-range planning 48
long-term orientation 21
long-term vision 106–107
loss of control 195
low cost competition 200
low tech innovation 212
low-technology 197
luxury goods 28

M

managing innovation 163–164
managing services 177
manufacturing jobs 64
margin erosion 79
market change 69, 113, 146–147
market conventions 126
market fit 115
market insight 148
market intelligence 213
market leadership 97
market opportunities matrix 148
market segmentation 37
market segments 28–29
market understanding 213
market-based innovation 145
marketing efficiency 38
marketing myopia 44
Marshall Industries 132–133
masculinity 21
mass consumption 1
mass marketing 37
mass screening 236
materialism 22–23, 27
materialists 22–23
McDonald's 38
mechanistic structure 162
medical devices market 216
medical insurance 58
medical solutions 85
me too propositions 33
Microsoft 143
middle classes 78
mission influence factors 113
mobile marketing 179
mobile technology 77
motivational domains 23

N

nano-manufacturing 203
national culture 19
natural gas 56
necessity goods 115
need identification 99
Netflix 186
network services logic 131
networking 48
networks 48, 154, 215
New Deal 104
new entrants 81
new product development 152
New Public Management 3, 229
new trade theory 60
niche products 201
non-renewable resources 60
Northern Rock 8, 40
Numatic 192
New Zealand police 232

O

O'Leary, M. 92
Obama, B. 106
obesity 85, 237–238
oil prices 5
oil 56
older consumers 246
Olsen, K. 135
online operations 177
online promotions 179
online retailing 101
online social couponing 78
OPEC 2, 37
open innovation 137–139, 168, 199
operational superiority 96
opportunity map 84
optical instruments 204
organic structure 162
organisational attributes 115
organisational learning 144
organisational structure 162
ostrich orientation 79–80
outsourcing innovation 167
own label products 39, 71

P

pacer technologies 147
pained-but-patient consumers 75
Pampers 32

participative decision making 167
people 177
Pepsi Cola 37
per capita income 12, 196
perishability 56, 151, 178, 218
Pilkington Glass 95–96
place 177
Polaroid Corporation 90
political economics 63–64
political environment 58–59
politicians 5
population ageing 5, 57
Porter's Diamond Model 61–62
positioning 31
post-materialism 22–23
post-materialists 22–23
power distance 21
pragmatic spenders 30
Pre-Depression Generation 25
premium brands 76
prevention innovation 234
price cuts 31
price decomposition 92–93
price points 33
price sensitive shopper 32
price–value relationship 48
process innovation 171, 198
Proctor & Gamble 32–33, 199
product commoditisation 65
product life cycle 121
product space map 114
product/market change matrix 150
product/service superiority 96
profit margin 15
Project Merlin 183
prospectors 119
prosperity trends 241
protestant work ethic 20–21
PSOs 226
PSO digitalisation 233
PSO entrepreneurship 228–229
PSO innovation 230
PSO innovation goals 233
PSO project managers 232
PSO technology 231
public sector debt 4
public sector deficit 226, 241
public sector 57
public sector spending 59
purchase motivation 27

purchasing behaviour 29, 71
Pyrex 111

Q
quality shopper 32
Quantitative Easing 8

R
R&D alliances 154
R&D co-operation 215
radical competence 95
radical change 91
radical innovation 32, 95, 136–137, 165
radical innovation hub 165RBV theory 99
reactors 119
re-adopting vision 111
reasoned purchasing 114
recession proofing 47
recession 69
Reckitt Benckiser 244
Red Ocean event 121
Redbox 187
regaining control 196
relationship marketing 42–43, 115
relative competences 90–91
reliability 180
religion 19
re-linking 101
resource based view (RBV) 87
resource flexibility 120
responsiveness 180
retail sales 29
retaining leadership 149
retirees 246
retrenchment strategy 150
Ricardo, D. 60
rivalry 61
Rolls Royce 206–207
Roosevelt, F.D. 104
Royal Bank of Scotland 41
Russia 14
Ryanair 92–93

S
sales promotions 31, 178
Samsung 196
scale of speed 197
Schumpeter, J. 16, 31, 45–46, 241

S-curve 156–157
sector exit costs 163
self image 27
semiconductor industry 16
semiconductor production 195
sensors 256
service efficiency 187
service in austerity 182–183
service demand 177
service effectiveness 187
service gap 180–181
service innovation 184
service provision 185
service quality 180–181
service scaling 99
service sector employment 64
SERVQUAL 180–181
shared values 162
shelf-life 155
Sherwood Electonics 192
short-term orientation 21
Siemens 14, 167
Silicon Valley 47–48
simplicity shopper 32
Singapore Airlines 88
slack resources 163
slam-on-the-brakes segment 75
small business lending 183
small firm innovation 219
small firms 166
small German firms 214
small restaurants 212
smart implants 256
smart strategy 253
smart technology 253
smart upwave 251–252
smart wave innovation 254
smartphones 77
SMEs 211
SME external knowledge 214
SME funding innovation 216
SME process innovation 220
SME radical innovation 217
SME survival 212
Smith, A. 60
SMS 179
social equality 59
social status 27
socio-demographics 29, 57
solar technology 242

Sonoco 51
Southwest Airlines 92–93
Spain 9, 248
Spectris 204
spider's web approach 49
stage gate model 152
state-of-the-art products 157
status consumption 27
status goods 74
strategic asymmetries 99
strategic capability 162
strategic complexity theory 120
strategic controls 98
strategic flexibility 118–119
strategic investment choices 98
strategic plan 49
strategic planning 48
strategic positioning 118
strategic response 49, 74–75
strategy sustainability 117
strategy 115–116
structural simplicity 121
student fees 80
sub-prime mortgage 6, 40, 182, 226
subscription websites 179
superior customer service 193
supermarkets 38–39
supply chain control 196
supply chains 50, 171
sustainable energy 242
sustained transformation 197
sustaining vision 108–109
symbolic management 104

T
tangibility 180
targeted promotions 178–179
task co-ordination 170
technical quality 180
technological change 142
technological competence 144
technological discontinuities 157
technological innovation 154
technological investment 151
technology brokers 199
technology fountains 199
technology isolationists 199
technology licences 65
technology opportunities 155
technology sponges 199

technology transfer 153
technology-based innovation 144–145
television advertising 37
Texas Instruments 158
Thatcher, M. 3
Thrift 245
tiered pricing 32
toxic debt 8
Toyota 94, 220
TQM 145
transactional marketing 43, 115
transformative capacity 168
trust 170

U
UK higher education 227
uncertainty avoidance 22
unemployment 2–3, 63
Unilever 33
unions 3
unique competence 100
Universities 79–80
upstream aggressors 82
US household furniture 200
US retail market 79
US Steel 56
USA 195
user know-how 171

V
value innovation 129–130
Value IQ Instrument 129
value migration 113

value-added jobs 64
values 22–23
venture capitalists 152
vertical networks 215
vision 83–84
visionary companies 108
visionary leaders 104–106
visionary political leadership 105
visioning 82–83
Volvo 219

W
Wal-Mart 88, 97
water 55
wealth generation 65–66
wealthy consumers 73
Weber, C. 20–21
welfare programmes 58
welfare state 1, 225, 241
Wessex Health Authority 232
Western consumers 69
Western democracies 58
wild card event 56
wind farms 243
wind technology 242

X
Xerox 135

Y
YouTube 222

Z
Zara 187

Printed and bound by CPI Group (UK) Ltd, Croydon, CR0 4YY

The manufacturer's authorised representative in the EU is Springer Nature Customer Service Centre GmbH, Europaplatz 3, 69115 Heidelberg, Germany. If you have any concerns regarding our products, please contact ProductSafety@springernature.com

Printed and bound by CPI Group (UK) Ltd, Croydon, CR0 4YY
23/03/2026
02076449-0012